SOCIAL CONSTRUCTIONISM IN HOUSING RESEARCH

Social Constructionism
in Housing Research

Edited by
KEITH JACOBS
University of Tasmania

JIM KEMENY
University of Uppsala

TONY MANZI
University of Westminster

Routledge
Taylor & Francis Group

LONDON AND NEW YORK

First published 2004 by Ashgate Publishing

2 Park Square, Milton Park, Abingdon, Oxon OX14 4RN
711 Third Avenue, New York, NY 10017, USA

Routledge is an imprint of the Taylor & Francis Group, an informa business

First issued in paperback 2017

British Library Cataloguing in Publication Data
Social Constructionism in housing research
    1.Housing - Research 2.Social psychology 3.Discourse
    analysis 4.Social interaction
1.Jacobs, Keith II.Kemeny, Jim III.Manzi, Tony

Library of Congress Cataloging-in-Publication Data
A catalog record for this book is available from the Library of Congress

ISBN 978-0-7546-3837-7 (hbk)
ISBN 978-1-138-27432-7 (pbk)

Contents

List of Figures and Tables

Figures

Tables

List of Contributors

David Clapham is Professor of Housing Studies at the Centre for Housing Management, Cardiff University.

Michael Darcy is Senior Lecturer in the School of Applied Social and Human Sciences, University of Western Sydney.

Anna Haworth is a Principal Lecturer in the School of the Built Environment and Architecture, University of Westminster.

Keith Jacobs is a Senior Lecturer in the School of Sociology, Social Work and Tourism, University of Tasmania.

Jim Kemeny is Professor of Housing and Urban Sociology at the Institute for Housing and Urban Research, University of Uppsala.

Peter King is Reader in Housing and Social Philosophy at the Centre for Comparative Housing Research, De Montfort University.

Tony Manzi is a Senior Lecturer in the School of the Built Environment and Architecture, University of Westminster.

Greg Marston is a Lecturer in the School of Social Work and Applied Human Sciences, University of Queensland.

Kathleen J. Mee is a Lecturer in Geography in the School of Environment and Life Sciences, University of Newcastle, New South Wales.

Max Travers is a Lecturer in the School of Sociology, Social Work and Tourism, University of Tasmania.

Preface and Acknowledgements

The idea for this book emerged from discussions following a recent joint paper written by the editors on the social construction of housing problems for the journal *Housing Studies*. It was apparent to us that in recent years housing researchers have been increasingly inspired by social constructionism. At the same time social constructionism has a long research tradition in the social sciences, stretching back to the late 1800s. Social constructionism also comprises a variety of perspectives – discourse analysis is just one of these – that many housing researchers are unaware of and that can provide a rich source of inspiration. Yet there are no books directly relevant to housing that bring together the range of different approaches deployed or provide a critical commentary on their application. We hope that this book, by showing the complexity and wide range of social constructionist perspectives and by providing examples of recent developments within the field, comprises a first step in filling this gap and in so doing will contribute to the development of a future social constructionist housing research agenda.

Our thanks to all of the contributors who responded to our requests with courtesy and delivered their papers in accordance to our time schedule. We are also grateful to Pat FitzGerald for her assistance in preparing the manuscript for publication.

Chapter 1

Introduction

Keith Jacobs, Jim Kemeny and Tony Manzi

The Return to Theory

One of the most encouraging features of contemporary housing research
has been a willingness to draw upon explicit theory from the different social
science disciplines. The best housing scholarship has always embraced social
theory: for example, political economy (Merritt, 1979; Ball, 1983; Maclennan,
1982; Muellbauer, 1990), political science (Dunleavy, 1981), social geography
(Harvey, 1973; Smith, 1989; Massey, 1994) and sociology (Rex and Moore,
1967; Castells, 1977; Saunders, 1990). However, as housing courses began to
be offered in the university sector in the 1980s and early 1990s, their growth
was not accompanied by significant explicit theoretical innovation. Instead
the development of the field of 'housing studies' was dominated by a strong
presumption that the primary task of research was to aid policy prescription.
Such an outcome can best be explained by the lack of enthusiasm from funding
bodies for theoretical innovation in favour of evidence-based policy analysis.

Though the commitment to prescriptive policy-based research continues, it
is no longer the case that theoretically explicit research is viewed as suspicious
or portrayed as superfluous. In fact, judging by the articles now appearing in
academic journals such as *Housing Studies*, *Housing Theory and Society* and
Urban Studies, there is an abundance of housing research embracing a range
of theoretical perspectives.

The rationale for this edited collection is to bring together in one volume
a detailed discussion of one theoretical perspective on housing, namely
social constructionism. We feel that such a volume is long overdue now
that social constructionism has emerged as an influential perspective and its
application extends to a number of different areas of housing research. This
introductory chapter discusses some general themes, all of which will be
developed further in subsequent chapters. It sets out the different influences
that have so far informed social constructionist perspectives in the field of
housing research; namely discourse theories, the sociology of social problems,
symbolic interactionism and the sociology of power. While to some degree

these themes overlap, it is expedient, nonetheless, to set out how each of these have informed contemporary housing research.

Before this, it is helpful to set out some of the reasons why housing academics have become more receptive to theory. Firstly, there is a now a widespread understanding that explicit theorising produces better quality research outcomes. In particular, it provides the basis for a clear framework that enables the reader to scrutinise research on its own terms and avoid ambiguity. Secondly, there is an appreciation that housing research cannot be undertaken successfully in isolation from wider debates taking place in other social science disciplines. Thirdly, the context of housing has altered significantly; in particular the residualisation of the British and Australian social rental housing stock and transformation of the nation state accentuated by globalisation processes have encouraged researchers to seek appropriate theories and concepts to make sense of these changes.

As mentioned above, there are also practical constraints upon theoretically-informed housing research, the most compelling of which relates to its funding and commissioning. For some years there has been disquiet that the specific demands of contract research and the pressure to generate income from consultancies has stifled opportunities to conduct critically orientated sociology. With some exceptions, the consultancies engaged in by housing academics focus on improving housing practice but limit the scope for explicit theory.

While the reasons above are applicable to all housing research which draws explicitly from social theory, there are specific factors that relate to social constructionism. Social constructionism is viewed by its adherents as providing a richer and more sociologically informed analysis of the policy process than traditional explanations. The impetus for much of the constructionist literature is an attempt by academics to develop a research agenda that is independent from the demands of policy makers. Arguably, a feature of constructionist research that distinguishes it from more traditional modes of analysis is the commitment to setting out an explicitly reflexive research methodology that makes clear the epistemological basis for investigation.

From the outset it is important to stress that social constructionism is just one of the social theories that has been utilised successfully by housing researchers to broaden the scope of housing research to cover issues that are not normally viewed as being of practical concern to policy makers. Nevertheless, constructionism functions as a counterweight to the dominance of the atheoretical empiricism of much traditional housing research. Additionally, it develops from the prior existence and more recent emergence of more explicit theories derived from positivist traditions.

What is Social Constructionism?

Since critics of social constructionism have claimed that it denies the existence of an objective material world, it is important from the outset to make clear that there is no attempt in this edited collection to advance such arguments. Instead, the claim advanced is that *our access* to the material world is mediated through language and discourse. In the words of Collin (1997: 2–3) 'our perception of the material world is affected by the way we *think* or *talk* about it, by our *consensus* about its nature, by the way we *explain* it to each other, and by the *concepts* we use to grasp it'. Of course the claims made by social constructionism raise a set of further issues. For example: what is meant by the idea that reality is mediated through language? In what sense can there be *deliberate* construction of social facts? Are there circumstances when groups of individuals can consciously impose their definition of social reality? What implications does this have for our ability to formulate issues, debate policies and to criticise or support government actions in the field of housing policy? These questions and others are discussed in the chapters that follow.

Since constructionism questions or problematises the notion of objective truth as such and instead emphasises the contingent basis of social reality, it is correctly portrayed as an antithesis of positivism. Rather than assuming that facts are given and discoverable through scientific investigation, constructionism questions the status of given assumptions and interrogates the process of 'claims-making' in social policy. Social facts are therefore understood as contingent, contested and subject to considerable diversity of interpretation. For housing academics writing from a constructionist perspective, research can never be simply a question of discovering facts and presenting them in a format amenable to policy makers. Nor can the cumulative weight of material evidence ever be sufficient to persuade the reader of the cogency of the analysis. The strength of constructionism is its focus on broader social processes and its emphasis on the importance of social, political and economic context.

Yet the constructionist perspective should not be construed as being simply a counterbalance to positivism or realism, for it also seeks to challenge the claims made in some of the more explicit theories that are derivative from positivism; for example, economic rationalism and behaviourism. Whilst these theories can enrich housing research by providing rigorous and enlightening perspectives, they can sometimes operate as a constraining influence; for example, effectively limiting the parameters of the subject matter to a discussion of the market or state's efficacy in addressing housing need. Social constructionism provides a means by which the subject matter of housing research can be extended beyond the confines of a 'state' versus 'market' narrative to cover areas generally perceived to be within the domain of cultural geography, ethnography and social anthropology.

It is necessary to provide a few words of explanation about the use of the term 'social'. Recently academics drawing from actor-network theories (e.g., Callon, 1998; Latour, 1999) have dropped the word 'social' altogether to indicate that our knowledge of the world is not just derivative of human interactions but of other interactions as well (e.g., technology, animals and the environment). However in this collection, readers will note that the terms 'constructionism' and 'social constructionism' are, on some occasions, used interchangeably. This is because the word 'social' is deployed in its widest sense; to denote that our understanding of reality is mediated through interaction – whether interpersonal, intrapersonal, person-technology, person-environment or any other kind.

The Manifold Strands of Social Constructionist Research

At this juncture it is helpful to set out the different strands of social constructionism that have been most influential in housing research.

Discourse Analysis

There is a common perception – at least in the housing variant – that constructionism is synonymous with discourse analysis. To date most of the constructionist studies in housing research have been derived from discourse perspectives that have their social science genesis in the work of Foucault (1980) and this has – up to now – been the most prevalent influence (e.g., Hastings, 1998; Gurney, 1999; Haworth and Manzi, 1999; Jacobs and Manzi, 1996; Jacobs, 1999; Kemeny, 2002b). Discourse analysis has been utilised because it provides a way of undertaking an evaluation of both text and the spoken word that can overcome some of the objections often made against non-positivist epistemologies.

In particular, the work of Fairclough (1995) has provided housing researchers with a framework for conducting policy-orientated research. Hastings' (2000: 133) explanation for the emergence of discourse analysis is that it provides a way of opening up 'new empirical terrain' or material previously ignored by researchers. Its utilisation by housing researchers can also be explained by the linguistic turn that influenced social sciences in the late 1980s and early 1990s (Fischer and Forrester, 1993). Although constructionist research in housing has been fuelled by an interest in language and its mediating effects, the disproportionate attention to discourse analysis in housing research represents a particular and late variant of constructionism with the result that other and earlier perspectives have been marginalised.

Social Problems and Policy Narratives

Work on the construction of social problems that has been particularly influential for housing academics includes Spector and Kitsuse (1977) and Schneider (1985). Rather than viewing a social problem as a given and objective condition, it is perceived as part of an interpretative process determining what should be seen as oppressive, immoral or intolerable. Thus issues such as poverty, crime or homelessness (Spector and Kitsuse, 1977) do not possess intrinsic qualities, but are seen as 'definitions of and orientations to *putative* conditions that are *argued* to be inherently immoral or unjust' (Holstein and Miller, 1993: 6, emphasis in original). This process of definition is both implicitly and intentionally rhetorical. Constructionism resists the 'essentialist' assumption that these problems have objective and identifiable foundations. Social problems are constructed on shifting sands of public rhetoric, coalition building, interest group lobbying and political expediency.

Social constructionist literature – for example Schattschneider (1960), Bachrach and Baratz (1962), Blumer (1971) and Spector and Kitsuse (1977) – maintains that agenda setting, advocacy coalitions, lobbying and media campaigns are all significant influences on shaping the policy process. As mentioned above, the rejection of 'taken for granted' perspectives, which view problems as simply reflections of underlying realities, allows instead explicit attention to be given to the factors that have to be in place before a housing problem becomes accepted and acted upon. These factors include the construction of narrative to tell a plausible story about a problem, the development of coalitions of support and the deployment of institutional resources to ensure a response.

The absence of constructionist housing problems in contemporary housing research is curious, as so much of the approach has been problem-oriented. Whilst discourse analysis became the entry point from which the majority of housing researchers began to take an interest in constructionism, as stated above, other variants have begun to attract the interest of housing academics.

For example, Sahlin (1995; 1996), Allen (1997), Jacobs et al. (1999; 2003) and Kemeny (2002a and b) draw upon a strand of constructionist literature that relates specifically to social problems and policy narratives. In particular they are interested in the role played by powerful interest groups in bringing housing problems into prominence through lobbying and policy making activity. So, for example, the significance of homelessness as a social problem is not only contingent on the material conditions as experienced by people in acute housing need but also on the ability of pressure groups to persuade policy makers that homelessness is an issue worthy of intervention.

Constructionist accounts can therefore illustrate the reasons behind changing approaches to social problems, helping to shed light on why particular issues assume prominence at different periods of time. They reject the notion of 'self-evident' problems, to be ascertained by a scientific process of discovery. Instead, such problems are intrinsically linked to the culmination of a complex process of bargaining, negotiation and struggles over definitions. These approaches lead to a wider focus not simply on government activity, but also on interest groups, media coverage, policy discourse and individual interpretations.

Symbolic Interactionism

A further strand of constructionism that has hitherto been largely overlooked by housing researchers is symbolic interactionism. The work of Berger and Luckmann (1961), Schutz (1967), Goffman (1959), Blumer (1969) and Garfinkel (1967) has been influential in a range of social policy applications including evidence based medicine, the sociology of deviance, organisational sociology, media studies and the sociology of science. Each of these writers stresses how aspects of our understanding are shaped by subjective experience and the mediation of these experiences in the context of social organisations and institutional settings (for example professional/client relationships in hospitals and schools).

Within the strand of constructionism inspired by symbolic interactionism, researchers emphasise social process rather than static structures. A priority of the research is to focus on the whole and to contextualise social situations, taking the definitions of actors as its starting point. Thus it explores the interactional basis of social organisation in different social contexts and at different levels of analysis. This strand of social constructionist literature stems from a concern that structural functionalism as expounded by Talcott Parsons (1951) was too abstracted from everyday life to be of use for empirically minded sociologists. Instead sociologists such as Goffman (1959) and Blumer (1969) emphasised the practicalities of everyday life and mapped out both its complexity and detail. The work of Strauss et al. (1964) is particularly relevant, for example their study of psychiatric institutions (1964) in which they deployed the concept of 'negotiated order' to explain the power relations between professional and patients in hospital settings. However as Kemeny (2002a: 141) points out, within housing research 'there has been an almost total absence of genuinely interactionist studies in the constructionist tradition'.

The Sociology of Power

The final strand of constructionism can be termed the sociology of power. Although this component of social theory is not exclusive to social constructionism many of the researchers interested in understanding the power dynamics within housing organisations have drawn upon a constructionist perspective to make explicit the network of power relations that affect the housing policy process. The foregrounding of competing actor and agency accounts of the policy process has resulted in some detailed sociological research (for example Sarre et al., 1989; Franklin and Clapham, 1997). The commitment to seeking out competing perspectives has meant that social constructionist accounts of the housing process seek to integrate individual and agency perspectives alongside wider structural concerns.

The work of Foucault (1980) has been especially influential in this respect, in particular his claim that power is not reducible to one single causal feature but is instead part of a network of relations. In making this claim Foucault casts aspersion on the formulation of power as a hierarchy, with the key actors within the state occupying the primary edifice: 'power is employed and exercised through a net like organisation. And not only do individuals circulate between its threads; they are always in a position of simultaneously undergoing and exercising this power' (Foucault, 1980: 98). Strategies to maintain control, according to Foucault, entail a range of processes including explicit and implicit methods for example discursive strategies and rhetoric. Such views have a clear relevance to analysing the politics of organisational relationships within the changing governance of social housing.

The dependence on sociological variants of constructionism indicates the selective and partial manner in which constructionism has so far influenced housing research. It also suggests that there is considerable potential for constructionist approaches from a range of disciplines – including political science, psychology, geography and even economics – to be applied to housing research. We hope that this collection will encourage the application of constructionism to housing research from a variety of disciplinary perspectives.

Structure of the Book

The book has been organised thematically. Chapters 2, 3 and 4 outline the major theoretical issues within the corpus, while Chapters 5 to 9 serve to demonstrate the applicability of social constructionism in different housing-related contexts (i.e., social exclusion discourses, individual housing choices, media production, organisational conflicts, and international practices).

The core argument within Chapter 2 by Max Travers, 'The Philosophical Assumptions of Constructionism', is that since epistemological presuppositions underpin all applied social research it is necessary to be as explicit as possible as to what these assumptions are. Travers argues that social constructionist research is primarily a reaction to the hegemony of positivism within the corpus of applied social research. As his chapter shows, the philosophical antecedents of constructionism include: the interpretive sociology of Max Weber; the phenomenological sociology of Alfred Schutz; the philosophy of science of Thomas Kuhn; and the linguistic orientated philosophies of Ludwig Wittgenstein and Peter Winch. Each of these writers, albeit from different vantage points, was concerned with how we experience social reality and how these experiences are mediated through interaction. Rather than seeing the sociological or philosophical enterprise as being one of foundation building that is developing scientific theories to explain human action, these writers sought instead to interrogate in close detail the ways in which we acquire and report knowledge. Travers also discusses the charge of 'ontological gerrymandering' made against social constructionism by writers who adopt a relativist and ethnomethodological perspective (Pollner, 1974; Woolgar and Pawluch, 1985; Pollner, 1991). They accuse social constructionists of not subjecting their own approach to the level of scrutiny that they apply to non-contingent methodologies such as positivism.

The following chapter, written by Peter King, seeks to build a case against social constructionist theories. We asked King to adopt an oppositional perspective and he dutifully obliged. King's chapter picks up on the general themes outlined by Travers but he concentrates his fire on how constructionist methodologies have been utilised by housing researchers. Particular targets include discourse analysis, which King contends is too ambitious both in its aims and approach, with the result that findings from this perspective are often rather obvious or banal. King also makes the charge of relativism although he acknowledges that his criticisms apply more to the strong version of constructionism. Nonetheless, for King even the weak version of constructionism is flawed because it seeks to apply a total explanation and purports immutable and absolute positions. Another concern of King is constructionist theories fail to acknowledge an embodied nature of selfhood and only explains selfhood in terms of social relations. For King this is inimical to believing that the individual subject is nothing more than an 'empty vessel filled up through discourse'. The concerns raised by Travers and King are important ones and feature in criticisms levelled against social constructionism by writers such as Ian Hacking (2001) but are not in our view insurmountable. The perspective adopted by the remaining authors in this book is that constructionism is not a grand all-embracing theory as King claims, but a theoretical perspective that has utility in certain research contexts but not in others. Likewise, there is nothing in the body of constructionist research committed to advancing a disembodied account of the self as King claims, though this said, there may be strong exponents of social constructionism who might maintain such a view.

The fourth chapter by Jim Kemeny, 'Extending Constructionist Social Problems to the Study of Housing Problems', makes use of the considerable body of literature on social problems including work by Merton (1951 and 1971), Mills (1959), Gusfield (1963) and Spector and Kitsuse (1973) to mount a defence of the constructionist position from different theoretical perspectives including objectivist, postmodern, Marxist and feminist critiques. Kemeny argues that housing researchers have largely ignored debates within the sociology of social problems literature and, though recently some writers have drawn inspiration from discourse-based methodologies, there remains considerable scope for housing researchers to modify their claims in response to recent criticism.

The next set of chapters take up Kemeny's challenge by examining a number of different housing and social policy issues from the perspective of social constructionism. For example, Greg Marston's chapter 'Constructing the Meaning of Social Exclusion as a Policy Metaphor' takes a critical look at the discourses surrounding the concept of social exclusion. Marston charts the ways in which social exclusion has been used by policy makers and argues for the need to critically interrogate the ways in which concepts such as social exclusion are deployed. His chapter has specific implications for policy practice, namely his observation that material deprivation continues to be the biggest obstacle undermining attempts by policy makers to foster civic participation and that policy prescriptions that overlook material factors are unlikely to make any significant impact.

David Clapham's chapter examines the concept of a 'housing pathway' that he himself has been instrumental in developing (Clapham, 2002). Clapham's model of the pathway entails applying this metaphor so as to explore the different housing circumstances that individuals encounter through their life course. He defines housing pathway as patterns of interaction (practices) concerning house and home over time and space. As Clapham points out, the pathway metaphor is deeply intertwined in the constructionist paradigm in its insistence that our understandings of what constitutes reality are to a considerable extent the outcome of interaction. It is a framework of analysis rather than a theoretical perspective and in this chapter he shows how it can illuminate how individuals make decisions and choices and how a whole range of different discourses can emerge out of this. For Clapham, the important dimensions for households lie in the relationship between work, housing costs, ideas about the home and neighbourhood and the meaning ascribed to each.

It is widely recognised that the media performs an instrumental role in setting the parameters of housing policy debate. In a detailed analytical study Kathleen J. Mee's chapter 'Necessary Welfare Measure or Policy Failure: Media Reports of Public Housing in Sydney in the 1990s', Mee charts how media perceptions towards public housing in Sydney changed over a 10 year period. Mee's analysis entails a combination of content and discourse analysis to show how public housing newsworthiness declined over the period and furthermore how reportage extenuated negative connotations of public housing, including stigmatisation.

Manzi and Darcy's chapter 'Organisational Research: Conflict and Power within UK and Australian Social Housing Organisations' explicitly uses a constructionist perspective to study conflicts taking place within housing organisations. Their chapter examines housing professional and organisational cultures and identifies a number of hegemonic discourses that set the context for policy development. These include a 'technical discourse' that has reinforced both centralised control and integration, 'commodification discourse' which strengthens market-based responses and 'democratisation discourses' which have valorised community or neighbourhood, including participatory mechanisms to engender tenant involvement in the management of housing. Manzi and Darcy's core argument is that it is the existence of these discourses that best explains the contradictions prevalent within housing management practices both in the UK and Australia.

Finally, 'Social Constructionism and International Comparative Housing Research' by Anna Haworth, Tony Manzi and Jim Kemeny examines the pitfalls associated with comparing different countries' housing systems from the perspective of the 'outsider'. As their chapter shows, the outsider faces many disadvantages in seeking to understand other countries' housing systems in sufficient depth. The most problematic is a tendency towards ethnocentrism. Yet, despite the challenges, comparative research has considerable value when it comes to making judgements about the efficacy of each nation's state housing system and for understanding how globalisation processes affect domestic housing policy agendas. Haworth, Kemeny and Manzi argue that constructionist methodologies, with their commitment to reflexivity and the importance that is attached to local context, offer a useful *modus operandi* to negotiate the obstacles that confront the housing researcher when engaging in comparative research. However, in spite of the tendency in academic research to emphasise globalisation and convergence of economic processes, this must never become a justification to prevent an understanding of the complexity and differences within each nation state.

As editors we hope that all of the chapters provide a useful source of reference for housing researchers and students who wish to draw upon constructionist methodologies. We also hope that even for those who wish to take issue with the arguments, the book provides a representative sample of what constructionist housing research entails, what is has to offer and the challenges that lie ahead.

References

Allen, C. (1997), 'The Policy and Implementation of the Housing Role in Community Care – A Constructionist Theoretical Perspective', *Housing Studies*, Vol. 12, No. 1: 85–110.

Bachrach, P. and Baratz, M. (1962), 'The Two Faces of Power', *American Political Science Review*, No. 56: 947–52.

Ball, M. (1983), *Housing and Economic Power: The Political Economy of Owner Occupation*, London: Methuen.

Berger, P. and Luckman, T. (1961), *The Social Construction of Reality*, London: Allen Lane.

Blumer, H. (1969), *Symbolic Interactionism: Perspective and Method*, Englewood Cliffs, NJ: Prentice Hall.

Blumer, H. (1971), 'Social Problems as Collective Behaviour', *Social Problems*, No. 18: 298–306.

Callon, M. (ed.) (1998), *Laws of the Market*, Oxford: Blackwell.

Castells, M. (1977), *The Urban Question*, London: Edward Arnold.

Collin, F. (1997), *Social Reality*, London: Routledge.

Clapham, D. (2002), 'Housing Pathways: A Postmodern Analytical Framework', *Housing Theory and Society*, Vol. 19, No. 2: 57–68.

Dunleavy, P. (1981), *The Politics of Mass Housing in Britain 1945–1975: A Study of Corporate Power and Professional Influence in the Welfare State*, Oxford: Clarendon Press.

Fairclough, N. (1995), *Critical Discourse Analysis: the Critical Study of Language*, London: Longman.

Fischer, F. and Forester, J. (1993), *The Argumentative Turn in Policy Analysis and Planning*, London: UCL Press.

Foucault, M. (1980), *Power/Knowledge: Selected Interviews and Other Writings 1972–1977*, ed. C. Gorman, Brighton: Harvester.

Franklin, B. and Clapham, D. (1997), 'The Social Construction of Housing Management', *Housing Studies*, Vol. 12, No. 1: 7–26.

Garfinkel, H. (1967), *Studies in Ethnomethodogy*, Englewood Cliffs, NJ: Prentice Hall.

Geertz, C. (1993), *The Interpretation of Cultures: Selected Essays*, London: Fontana.

Goffman, E. (1959), *The Presentation of Self in Everyday Life*, Harmondsworth: Penguin.

Goffman, E. (1967), *Interaction Ritual*, New York: Anchor.

Gurney, C. (1999), 'Pride and Prejudice: Discourses of Normalisation in Public and Private Accounts of Home Ownership', *Housing Studies*, Vol. 14, No. 2: 163–83.

Gusfield, J. (1963), *Symbolic Crusade: Status Politics and the American Temperance Movement*, Urbana: University of Illinois Press.

Hacking, I. (2001), *The Social Construction of What?*, Boston: Harvard University Press.

Harvey, D. (1973), *Social Justice and the City*, London: Edward Arnold.

Hastings, A. (1998), 'Connecting Linguistic Structures and Social Practices: A Discursive Approach to Social Policy Analysis', *Journal of Social Policy*, No. 27: 191–211.

Hastings, A. (2000), 'Discourse Analysis: What Does it Offer Housing Studies?', *Housing Theory and Society*, Vol. 17: 131–9.

Haworth, A. and Manzi, T. (1999), 'Managing the "Underclass": Interpreting the Moral Discourse of Housing Management', *Urban Studies*, Vol. 36, No. 1: 153–65.

Holstein, J. and Miller, G. (1993), 'Reconsidering Social Constructionism', in J. Holstein and G. Miller (eds), *Reconsidering Social Constructionism: Debates in Social Problems Theory*, New York: Aldine de Gruyter: 5–23.

Jacobs, K. (1999), 'Key Themes and Future Prospects: Conclusion to the Special Issue: Discourse and Urban Change' (editorial), *Urban Studies*, Vol. 36, No. 1: 203–13.

Jacobs, K. and Manzi, T. (1996), 'Discourse and Policy Change: The Significance of Language for Housing Research', *Housing Studies*, Vol. 11, No. 4: 543–60.

Jacobs, K. and Manzi, T. (2000), 'Evaluating the Social Constructionist Paradigm in Housing Research', *Housing Theory and Society*, Vol. 17, No. 1: 35–42.

Jacobs, K., Kemeny, J. and Manzi, T. (1999), 'The Struggle to Define Homelessness: A Constructivist Approach', in D. Clapham and S. Hutson (eds), *Homelessness: Public Policies and Private Troubles*, London: Cassell: 11–28.

Jacobs, K., Kemeny, J. and Manzi, T. (2003), 'Power, Discursive Space and Institutional Practices in the Construction of Housing Problems', *Housing Studies*, Vol. 18, No. 4: 429–46.

Kemeny, J. (2002a), 'Reinventing the Wheel? The Interactional Basis of Social Constructionism', *Housing, Theory and Society*, Vol. 19, Nos 3–4: 140–41.

Kemeny, J. (2002b), 'Society Versus the State', *Housing, Theory and Society*, Vol. 19, Nos 3–4: 185–95.

Latour, B. (1999), 'On Recalling ANT', in J. Law and J. Hassard (eds), *Actor Network Theory and After*, Oxford: Blackwell.

Maclennan, D. (1982), *Housing Economics: An Applied Approach*, London: Longman.

Massey, D. (1994), *Space, Place and Gender*, Cambridge: Polity Press.

Merritt, S. (1979), *State Housing in Britain*, London: Routledge and Kegan Paul.

Merton, R.K. (1951), *Social Theory and Social Structure*, Glencoe: Free Press.

Merton, R.K. (1971), 'Epilogue: Social Problems and Sociological Theory', in R.K. Merton and R. Nisbet (eds), *Contemporary Social Problems*, New York: Harcourt, Brace, Jovanovich: 793–864.

Mills, C. Wright (1959), *The Sociological Imagination*, New York: Oxford University Press.

Muellbauer, J. (1990), *The Great British Housing Disaster and Economic Policy*, London: Institute for Public Policy Research.

Parsons, T. (1951), *The Social System*, New York: The Free Press.

Pollner, M. (1974), 'Sociological and Common-sense Models of the Labelling Process', in R. Turner (ed.), *Ethnomethodology*, Penguin, Harmondsworth: 27–40.

Pollner, M. (1991), 'Left of Ethnomethodology', *American Sociological Review*, Vol. 56: 370–80.

Rex, J. and Moore, R. (1967), *Race, Community and Conflict*, Oxford: Oxford University Press.

Sahlin, I. (1995), 'Strategies for Exclusion from Social Housing', *Housing Studies*, Vol. 10, No. 3: 381–401.

Sahlin, I. (1996), 'From Deficient Planning to "Incapable Tenants". Changing Discourses on Housing Problems in Sweden', *Scandinavian Housing and Planning Research*, Vol. 13, No. 4: 167–81.

Sarre, P., Phillips, D. and Skellington, R. (1989), *Ethnic Minority Housing: Explanations and Policies*, Aldershot: Avebury Press.

Saunders, P. (1990), *A Nation of Home Owners*, London: Unwin Hyman.

Schattschneider, E. (1960), *The Semi-Sovereign People*, New York: Holt, Rinehart, and Winston.

Schneider, J. (1985), 'Social Problems Theory', *Annual Review of Sociology*, 11: 209–29.

Schutz, A. (1967), *The Phenomenology of the Social World*, Chicago: Northwestern Press.

Smith, S. (1989), *The Politics of Race and Residence*, Cambridge: Polity Press.

Spector, M. and Kitsuse, J. (1973), 'Social Problems: A Re-formulation', *Social Problems*, 21: 145–59.

Spector, M. and Kitsuse, J. (1977), *Constructing Social Problems*, New York: Aldine de Gruyter.

Strauss, A., Schatzman, L., Ehrlich, D., Bucher, R. and Sabshin, M. (1964), *Psychiatric Institutions and Ideologies*, New York: Free Press.

Woolgar, S. and Pawluch, D. (1985), 'Ontological Gerrymandering', *Social Problems*, Vol. 32: 214–27.

Chapter 2

The Philosophical Assumptions of Constructionism

Max Travers

Since the publication of Berger and Luckmann's (1967) *The Social Construction of Reality*, constructionism has become an influential movement across the human sciences, and in sociology has generated productive empirical programmes in the subfields of the sociology of science and social problems research. It has also generated lively philosophical debates about the extent to which sociologists can obtain objective knowledge about the world, and the problem of relativism. My aim in this chapter is to explain what is at issue in these debates, and to show how they are relevant to constructionist research in the field of housing studies.

The chapter will begin by explaining why philosophical debates are important for sociology, even if they are not always acknowledged. It will focus on the debate between realist and interpretive traditions and review the anti-realist arguments of Alfred Schutz, which influenced Berger and Luckman. The next section will look at the issue of relativism through considering the ideas of Peter Winch and Thomas Kuhn. It will then review how these issues are relevant to sociology through examining the field of social problems research, a subtradition of symbolic interactionism which has investigated how these are 'socially constructed' through the claims made by politicians, the media and pressure groups.

This part of the chapter will explain how researchers have come to spend increasing amounts of time debating epistemological issues, focusing on the critiques advanced within this field by poststructuralists and ethnomethodologists. However, it will also show how these philosophically driven critiques have methodological implications, so there are a variety of ways of studying social problems from a constructionist perspective. The chapter will conclude by considering the relevance of these debates and approaches for housing researchers, and possible ways in which constructionist research in this field could develop.

The Importance of Philosophy

Peter Winch (1988) once described sociology as 'misbegotten epistemology', meaning that even the most empirically-grounded study – what today one would call evidence-based policy – rests on unstated philosophical assumptions. The tradition that has remained dominant, although not without facing considerable criticism in the academy, is realism. There are many varieties that include several versions of positivism (Halfpenny, 1982), and critical realism (see for example, Bhaskar, 1978; Sayer, 1984). They all believe that the social world has an objective and independent existence in the same way as nature, and can be studied using scientific methods.

Comte (1974) and Durkheim (1966) established positivism in sociology as a method that involved the careful description of 'social facts', and the identification of structures and laws that shaped and determined human behaviour. They argued that this approach made it possible to devise more effective policy interventions in addressing the social problems associated with industrialisation. The statistical techniques used by Durkheim were developed by Paul Lazarsfeld and his colleagues at the University of Columbia in the 1940s into the sophisticated quantitative methods, such as multivariate analysis, that are used by contemporary sociologists (see Lazarsfeld and Rosenberg, 1955). Although they have little interest in debates about methodology, one can see how researchers who conduct empiricist studies for government agencies in applied fields like housing studies have similar epistemological assumptions. Policy analysts also believe that social processes can be measured and described, findings presented and recommendations made which can ameliorate, if not solve, social problems.

Positivism as a philosophy of social science was criticised in the post-war period for encouraging political quietism or conservatism among academics. Drawing on Marx's writings, and the critique of positivism made by the Frankfurt School, left-wing researchers developed what has become known as critical realism as an alternative epistemological position (see for example, Bhaskar, 1978; Sayer, 1984). In contrast to the search for causal laws, this places more emphasis on identifying the mechanisms that can explain the relationship between the variables, such as those identified in Marx's theory of historical materialism.[1] It is worth noting that while a lot divides positivists and critical realists politically, they both believe that one can explain human behaviour through identifying structures or laws that are hidden from ordinary members of society.

The opposing epistemological position to realism is interpretivism, the view advanced by Weber and other nineteenth-century German thinkers that social science requires different methods and assumptions to natural science, and must address the meaningful character of human group life. From this perspective there are no independent separately-existing structures that determine human action: everything is constructed or constituted through our ability to understand and give meaning to objects in the world and other peoples' actions.[2]

Weber (1978: 1) saw the task of interpretive sociology as understanding 'all human behaviour when and in so far as the acting individual attaches a subjective meaning to it' through using a method he called *verstehen* or empathic understanding. This enables the analyst to recognise a woodcutter in a forest, but also to establish whether he is 'working for a wage' or perhaps 'working off' a 'fit of anger' (Weber 1978: 3). Weber argued that, although this makes sociology more difficult than natural science, it also makes it more rewarding. There is far more to investigate and appreciate in human cultural life once one gives up the inappropriate natural science model.

It was, however, Alfred Schutz – a German refugee writing in America after the Second World War, and influenced by the phenomenological philosopher Edmund Husserl – who pursued the question of meaning more systematically through identifying a number of problems and ambiguities in the way Weber understood the concepts of action and *verstehen*. Schutz noted, for example, that it is impossible to access subjective meaning, but that one can immediately see what a woodcutter is doing through drawing on 'intersubjective' or public knowledge (1967: 32), shared in the form of typifications acquired in the course of socialisation.

Much of Schutz's work attempts to explain how intersubjectivity works through investigating the 'lifeworld' of our ordinary, everyday understandings. From this perspective, the positivists and even Weber had taken the existence of an objective social world for granted, whereas how people ordinarily experience this world could be investigated through reflecting on the nature of consciousness. Schutz was struck by how we are born into, and have to find our way around, a world that has an independent, objective existence:

> I find myself in my everyday life within a world not of my own making … I was born into a preorganised social world which will survive me, a world shared from the outset by fellow men who are organised into groups (Schutz, 1973: 329).

Whereas Durkheim believed that this could be studied by establishing a causal relationship between objective 'social facts', Schutz was interested in how this constraint is actually experienced and understood in everyday life. He was, for example, interested in how typifications and recipe knowledge are employed in making sense of situations and dealing with practical circumstances, and how new knowledge and skills can be created in response to unforeseen events. He asked how we can understand the actions and motives of others, and investigated the distinctive character of different 'multiple realities': how one experiences the world in a dream; as a scientist pursuing a rational systematic project; and in everyday life. In a similar way to Weber, he argued that one cannot address the meaningful character of life in its full complexity, but that any sociological explanation that uses generalised schemes or models to represent action should respect the understandings that people actually do have in the life-world.

This last point has considerable implications for sociological method in that positivists like Durkheim were committed to devising an objective scientific language that was superior to ordinary commonsense knowledge and saw no need to investigate this empirically. This radical difference in philosophical assumptions also explains why Parsons and Schutz were unable to develop a fruitful dialogue when they exchanged letters (Grathoff, 1978). Here one might note that Berger and Luckmann's (1967) integration of Schutz with Parsons, although successful in popularising his ideas to a wider audience, was in some respects misconceived since it glossed over these epistemological differences. In common with many subsequent attempts to solve the 'micro–macro' or 'action–structure' problem in terms largely set by Parsons, *The Social Construction of Reality* tends to favour the perspective of the objective, scientific analyst rather than looking at how ordinary people understand the practical circumstances of their own lives.[3]

One can also see that, however much describing reality as a 'social construction' appealed, and still appeals, to sociologists as a political slogan directed against aspects of society one would like to change, it misrepresents Schutz's philosophical position. Here one has to speculate since Schutz died in 1959, but it is interesting that he never used the term 'construction' in his own work. This is because he was interested in showing how the world is experienced as objective, as something that cannot easily be changed. This has some similarities with Max Weber's appreciation of how politics involves 'a strong and slow boring of hard boards' (Weber, 1991: 128). Berger and Luckman, by contrast, offer a considerably more optimistic view of construction, which suggests that society can be changed without too much difficulty.[4]

The Issue of Relativism

Before looking at how relativism has become an issue for sociologists undertaking constructionist research, it is important to appreciate how it is understood and debated by philosophers as a purely intellectual pursuit. This is important since arguments from philosophers like Wittgenstein and Kuhn, or interpretations of their ideas, are employed by many sociologists, although there are others who dislike the philosophical character of these arguments, and have called for a return to honest description and analysis (see Best, 1993).

The view that there is no such thing as absolute truth and no firm or objective foundations to knowledge has been advanced by a number of intellectual traditions and thinkers since the Enlightenment period. The most celebrated recent examples would be Michel Foucault (1972), who saw all ideas as the product of arbitrary and contingent historical processes, and Jacques Derrida (1976), who argues, among other things, that there is no fixed meaning to any text (which would include the reports produced for government agencies by housing researchers).[5] One can also see, however, that the interpretive stance of taking meaning seriously, promoted by many German thinkers in the late nineteenth century, including Dilthey, Gadamer and Weber, as an alternative to positivism, raises questions about relativism, since groups and individuals understand the world in different ways.

Rather than considering the whole tradition, I will be concentrating on the debates that took place in Britain during the post-war period concerning the ideas of Peter Winch and Thomas Kuhn. These are instructive in the sense that they give a taste of what philosophers, as opposed to sociologists, mean by the issue of relativism. It also shows how there are usually disagreements among philosophers, to an even greater extent than in sociology, over what any particular thinker actually believes. Winch and Kuhn have been vilified and applauded for being relativists, while denying that this is a correct reading of their work.

Peter Winch and Relativism

Winch (1988; 1977) provides a similar set of philosophical arguments to Schutz for what is wrong with positivism, and for how to pursue an interpretive social science. While Schutz drew on phenomenology in reflecting on consciousness, Winch applied the ideas developed by Wittgenstein (1958) in the *Philosophical Investigations* to understanding social science. The principal argument is that a scientific approach to studying human beings is inappropriate and misguided, since our lives are meaningful; not an original argument but one directed with considerable force and precision against positivism.

Wittgenstein's central argument against philosophy, whether positivist or otherwise, is that many of the problems it claimed to address were, in fact, pseudo-problems, and that this could be shown to be the case through paying attention to ordinary language. Relativism would be typical of the problems created by sceptical philosophers, even though it is not a problem in everyday life.[6] Winch argues that sociological investigation should involve understanding the rules that constitute different areas of life rather than attempting to construct a scientific theory explaining human action, the procedure adopted by positivist thinkers like Durkheim. In fact, he argues that the concept of causation is inappropriate in understanding human action: people have reasons for their actions, and without this insight, one would only produce a thin and meaningless account of human behaviour.

He took this argument against positivism a step further by questioning the assumption that scientific rationality could be used as a standard to assess the practices of a different culture. The pretext was a passage by the anthropologist Evans-Pritchard characterising Azande witchcraft beliefs as if they were a deficient version of Western science. This was widely perceived as an argument for relativism: for suggesting that there was no basis for saying their beliefs about the effectiveness of magic were false; and that it was impossible to understand another culture. To put this another way, everything is constructed through different 'language games' that develop contingently, and could just as well be different: there is consequently no secure basis in a philosophical sense for making an evaluative judgment.

In fact, Winch was making the less contentious argument that each 'form of life' or 'language game' has to be understood in its own terms, and if one does this one obtains a much richer and more complex understanding of society than by asking positivist questions that involve testing hypotheses. Understanding another culture may be difficult, but it is not impossible, and does not require sophisticated technical methods. Moreover, the fact that social life develops historically, and one speaks from a particular cultural location, is not viewed as a disturbing discovery, but as something everyone knows about, and which could hardly be different. Despite the fact that they are often viewed as relativist constructionists, Winch and Wittgenstein are as committed to the idea that we inhabit an objective social world as is Schutz, although they ground this in our shared understanding and mastery of language.

Thomas Kuhn and Relativism

The philosopher who has created most controversy over his alleged relativism is Thomas Kuhn, partly because he appeared to be challenging the authority of natural science. It is interesting that what are, to some extent, standard sociological arguments about how institutions have developed contingently over time, and could theoretically have taken a different form, no longer provoke fierce denunciation from politicians or religious leaders (although this was the case during the Enlightenment period) but can produce a reaction from some scientists. This also explains the lively character of the sociology of science, although it should be noted that, despite the best efforts of practitioners who have included ethnographers, ethnomethodologists, hyper-relativists, actor-network theorists, and various interpreters of Wittgenstein, it has still only had a limited impact on the mainstream discipline.

Kuhn (1996) challenged the received philosophical view of how scientific knowledge develops steadily through history as theories are falsified when new experimental information is obtained about the natural world. Instead, he argued that change occurred during revolutionary periods when a new world view or paradigm was adopted. It was often unclear until years later that the new paradigm was superior, and although it succeeded politically in obtaining control of intellectual associations and institutions, most scientists brought up within the old paradigm were unable to understand or accept the new way of thinking.

Kuhn also made some radical observations about the relationship between language and reality. Many scientists, as well as ordinary members of the public, believe that nature exists separately to how we describe it, and so scientific progress involves having our theories checked by nature. By contrast, Kuhn argued that it can equally well be described and understood in many ways: there is no theory-independent world, and so it makes no philosophical sense to argue that a scientific revolution results in a better understanding of nature.

This view was enthusiastically taken up in the sociology of science, and was pursued using a combination of historical, ethnographic, and discourse analytic methods.[7] To give one example, what became known as the 'strong programme' advanced the philosophically provocative view, through a number of historical studies, that the reception given to any scientific theory can be explained in terms of social factors or interests (see for example, Bloor, 1991; Bloor and Barnes, 1992). A key methodological principle was that explanations had to be 'symmetrical'; even theories that have become accepted as 'true' accounts of nature are social constructions.

Hughes and Sharrock summarise the relativistic assumptions of this research programme in the following terms:

> One implication of such a view is to render meaningless the quest for intellectual authority, as positivism did, for example, through a philosophically secure conception of the foundations of human knowledge. Philosophy too, as a body of knowledge, is socially caused and, hence, dependent on the social conditions which produced it. It is social conditions that determine what will be accepted as knowledge ... not any absolute principles or criteria independent of social determination. Accordingly, there are no secure foundations for human knowledge: all knowledge is relative (Hughes and Sharrock, 1997: 84–5).

This sounds like a radical statement, and one can see how the strong programme is potentially subversive in discovering and promoting 'suppressed' or minority views against scientific orthodoxy. Nevertheless, to a large extent this misses the point of what motivates researchers in this field. What most interests constructionists writing about science is how one might best understand the relationship between language and reality philosophically, and the debates have almost no interest or purpose outside the realm of academic philosophy. This is also the case for deconstructionism and other relativist programmes in the human sciences. They sometimes provoke an outraged reaction from defenders of institutions like science, literature and law, without actually challenging these institutions.[8] This will also be apparent in the next section, where I examine how what are essentially philosophical debates about constructionism have become increasingly important in the field of social problems research.

Philosophical Debates in Social Problems Research

The field of social problems research grew out of the labelling tradition founded by interactionist sociologists.[9] Howard Becker, Edwin Lemert and others proposed that instead of treating deviance as an objective social fact, attention should be focused on the interpretive procedures employed in recognising deviance. This was politically subversive in that labelling theorists advocated the decriminalisation of cannabis, and questioned the existence of mental illness. It could not, however, survive against a resurgent Marxism that appeared to offer a scientific means of understanding the real causes of deviance that were hidden from ordinary members of society.

Social problems research has been less political than the labelling tradition in the sense that it is no longer associated with a counter-cultural movement seeking to transform American society. On the other hand, it has drawn attention to how powerful interest groups such as the tobacco industry promote their own interests, and the difficulties facing new social movements in changing public attitudes. It has not had to face much criticism from Marxists or critical theorists (although see Agger, 1993), since there are few around in the contemporary academy, and they are still coming to terms with the triumph of liberal capitalism following the collapse of the Soviet Union in 1989. There have, however, been two other theoretical objections, each of which has a philosophical basis, in that they point to logical inconsistencies in social problems research. In the rest of this chapter, I will explain the basis of these arguments, but also show how they have implications for how one conducts empirical research.

The Relativist Critique

The most telling, not to say devastating, critique of social problems research (in the sense that everyone in this field acknowledges the truth of the criticism) has been made by relativists who are unhappy with how researchers use the term 'construction'. In simple terms, the objection is that it is usually used as a means of undermining arguments or institutions one disagrees with on moral or political grounds. These are 'constructions', and constructionist studies often demonstrate how they arise through interests or contingent social processes, and make the familiar sociological claim that things could be different. On the other hand, the positions one agrees with, and the 'facts' that support these arguments are not subjected to the same treatment.

Woolgar and Pawluch (1985) note that the typical social constructionist study involved identifying some constant condition and then examining how claims are made about it by different groups. These claims are presented as 'constructions', but what is not acknowledged is that the constant condition is exempt from this treatment. They give examples of studies about the 'discovery' of child abuse but which also reveal 'a commitment to the realist view of child-abuse' (Woolgar and Pawluch, 1985: 221). This would not be a problem if researchers admitted this was the case, and if they were pursuing a realist form of explanation. Instead, however, they were committed to a constructionist (or in the terminology used by labelling theorists) 'constitutive' approach to understanding social problems.

Woolgar and Pawluch were particularly interested in the method used to pursue this logically contradictory form of analysis in sociological texts. They argued that this was accomplished through a rhetorical strategy called 'ontological gerrymandering':

> But how do authors manage to portray statements about conditions and behaviours as objective while relativizing the definitions and claims made about them? The metaphor of ontological gerrymandering suggests the central strategy for accomplishing this move. The successful social problems explanation depends on making problematic the truth status of certain states of affairs selected for analysis and explanation, while backgrounding or minimising the possibility that the same problems apply to the assumptions upon which the analysis depends (Woolgar and Pawluch, 1985: 216).

In some respects, this is a version of the 'symmetry' argument in the strong programme of the sociology of science applied to the study of social problems. Everything in the world can be problematised as a 'construction', including the analyst's own position in producing a supposedly objective overview of the interest groups and social processes producing the constructions. The response of most constructionists was to accept both the force of the argument, and that there was a distinction between 'strong' and 'weak' constructionism. They were guilty of 'ontological gerrymandering' as charged.

The Ethnomethodological Critique

Criticism of a different kind was directed at the field of social problems research by ethnomethodologists. Here it should be noted that Harold Garfinkel, the founder of this sociological approach, was influenced initially by Schutz, and his ideas are also close to those of the later Wittgenstein in emphasising the importance of language as constituting reality. In reviewing their ideas earlier in the chapter, I noted that they appreciated the objective and constraining character of the world for ordinary members of society, and would have been uncomfortable with the charge of being 'relativists'.[10] Garfinkel also took issue with how the term 'construction' was used in the interactionist tradition, even before labelling theory. He complained that Goffman's dramaturgical observations about the production of gender did not do justice to how people experienced this as an objective and constraining feature of their lives (Garfinkel, 1967: 165–7).

An early ethnomethodological critique against social problems research was made by Melvin Pollner (1974) when it was still known as labelling theory. He had conducted ethnographic research about how disputes were resolved in the lower criminal courts, and was eventually led to a relativist position in interpreting this data (Pollner, 1991). However, in this paper he undermines the potentially relativist implications of constructionism by focusing on another logical problem. Becker and others had proposed a programme which would identify how deviants were produced ('constructed') by agencies such as the police and the courts. However, in formulating a research agenda, he recognised that there could be 'secret deviants' who had not yet been labelled. Pollner's objection was that there could logically be no such thing as a 'secret deviant': deviance was constituted by the interpretive processes that make up labelling.[11]

This debate raises in a pointed way a central issue or problem within interpretive sociology. If one follows Schutz, Wittgenstein or Winch, the only task for the sociologist is to describe how members of society understand their own activities. The concept of a 'secret deviant' suggests that one can do more than this; but this is only possible by what Woolgar and Pawluch call 'ontological gerrymandering', by claiming that you have independent access to an objective reality. The ethnomethodological response is that there is no independent access, but rather than opening the Pandora's Box of relativism, it invites an appreciation of how people understand social problems.

Bogen and Lynch (1993), drawing on Wittgenstein to pursue this line of argument, took issue with interactionists for artificially using the concept of 'social problems' to create a sociological field even though social actors do not normally understand their actions in these terms. They note, for example, that:

> The status of crime as a social problem for the most part is not thematic to the various language games in criminal justice institutions, nor is it thematized in studies of these particular language games. It would be unusual to find, for instance, adversaries in a criminal trial debating about whether or not 'common assault' is a 'social problem'. They would more likely argue about whether or not a particular defendant committed the assault in question, whether his 'joining a fight' in fact constituted 'an assault', or whether, having been convicted, the assailant remains a 'menace to society' (Bogen and Lynch, 1993: 227).

This shifts attention from the study of social problems, as for example they are presented in the media, to looking at 'the complex fields of practical action in which social problems arise' (Bogen and Lynch, 1993: 231). Other ethnomethodologists have argued that instead of examining and reproducing political arguments about gender, it would be more fruitful to understand how gender is produced interactionally in everyday life (Marlaire and Maynard, 1993). These can be seen as attempts to broaden the field by introducing new methods and topics. But they can also be viewed as being motivated by a desire to challenge the residual realist assumptions that still underpin interactionist research. As Lynch and Bogen argue, the very term 'social problems' research may actually prevent an appreciation, or detailed investigation, of different social worlds.

Methodological Considerations

One additional point worth making about these debates is that, although they were about philosophical issues, they also had methodological implications.[12] From an ethnomethodological perspective, the interactionist programme on deviance or social problems did not go far enough in addressing how labels were actually applied or social problems recognised. One could argue that they were too quick to take objective or structural conditions for granted instead of investigating how social actors understood these as part of the topic. A related criticism from conversation analysts was that one could investigate the world in more detail by studying naturally-occurring conversation.

The relativist critique was also not simply an exercise in raising philosophical difficulties. Some analysts, drawing on initiatives that had already taken place in anthropology, saw this debate as an opportunity to introduce new forms of postmodern ethnographic writing into the field of social problems research (see Holstein and Miller, 1993: chs 17–23). The complaint here was that traditional constructionist studies – even those which contrasted different perspectives and versions – still suggested that it was possible to write about the world objectively. An alternative style of writing, developed by Woolgar and others, was intended to challenge this assumption by drawing attention to how any study was itself a construction, and was only producing a version of reality. This was done by, for example, introducing imaginary 'playful' dialogues into what started as a scientific review of the literature, and framing the analysis of ethnographic or historical data with extended hyper-reflexive discussion of epistemological issues.[13]

Finally, there was a 'fight-back' from traditional social problems researchers who complained that engagement with philosophical issues and arguments had damaged this field. While acknowledging that there was no answer to the philosophical objections of the relativists, Joel Best complained that this diverted researchers from doing empirical research:

> Just as quantitative researchers continually risk sacrificing sociological substance for more elaborate research designs and more sophisticated statistics, qualitative researchers must balance substance against the demands of theoretical consistency. Analytic purity can come at a terrible cost … The sociology of

social problems began with the assumption that sociological knowledge might help people understand and improve the world; strict constructionism sells that birthright for a mess of epistemology (Best, 1993: 143).

Best recommended that researchers turn away from the attractions of postmodern ethnography, and instead develop 'grounded theories through analytic induction' (Best, 1993: 144). It should be noted that this tradition is informed to some extent by positivist assumptions, so one can see how even a return to 'honest' research, uncomplicated by philosophical considerations, can rest on concealed epistemological assumptions.

The Implications for Housing Studies

This chapter has attempted to explain some of the debates between realists and interpretivists that one can find discussed at greater length in texts on the philosophy of social science, and to show how they have been relevant to the sociology of science and the field of social problems research. Since there are probably housing researchers, and not just empiricists, who will share Joel Best's view that too much engagement with philosophical questions can divert sociologists away from their main task of helping 'people understand and improve the world', it is worth ending this chapter by considering the implications for housing studies (even though I have not contributed to this field, and would not consider myself an expert). What is the relevance of philosophical debates about realism and interpretivism for housing research? Why should housing researchers pay attention to the philosophical debates surrounding constructionism? And to what extent should they be concerned about the spectre of relativism?

Constructionist research on housing has attempted to develop an explicitly theoretical intellectual programme in a field where most research is funded by government departments or agencies committed to what, in terms of this chapter, is a positivist or empiricist understanding of evidence-based research (Jacobs and Manzi, 2000). It has so far drawn on a range of theoretical resources, including the social problems tradition in symbolic interactionism reviewed in this chapter, but also structuration theory, critical discourse analysis and Foucauldian network theory.

Although most discussion in this field has been concerned with developing and justifying this new approach as an alternative to empiricism, it should be apparent from this chapter that there are all kinds of debates within this alternative paradigm. The central issue has always been a concern about relativism, and I have tried to show how this arises from the interpretive interest in the meaningful character of human group life. Philosophers who have pursued this line of thought are often accused of being relativists, and I have argued that Schutz, Wittgenstein, Winch and Kuhn were, in fact, not attempting to challenge or undermine our ordinary experience of an objective world. That said, relativist philosophical arguments have been introduced into constructionist fields like social problems research and the sociology of science, and have generated much discussion and argument.

It would be fair to say that none of the epistemological debates reviewed in this chapter, either between realists (who include positivists like Comte and Durkheim, and critical realists like Bhaskar and Sayer) and interpretivists, or between relativist and their opponents in the interpretivist camp, have so far informed, or been pursued, in a thorough-going way by housing researchers. This is partly because, as Woolgar and Pawluch demonstrated in relation to social problems research, it is in practice easy to combine different epistemological assumptions, or shift between different theories when conducting a piece of sociological analysis. Most British sociologists are not greatly troubled by epistemological issues, whether these have been raised by interpretivists or poststructuralists. There is a general recognition that both 'action' and 'structure' are important, and can be addressed without great conceptual difficulty when conducting empirical research.

As this field develops, one would expect to find more debate between researchers with different epistemological or theoretical commitments. There is, for example, a potential tension between discourse analytic studies grounded in a realist understanding of the structures of contemporary capitalism, and a Foucauldian approach that treats all theories as discursive constructions. Although this was acknowledged by Annette Hastings in her introduction to the 1997 special issue of *Urban Studies* on discourse analysis and urban change, most contributors did not make an issue of epistemological debates between Foucault and Marx, or their implications for political practice.

There is also a potential tension between realist and interpretive understandings of constructionism. Although they have often slipped back into realism in studying social problems, symbolic interactionists are committed to addressing the 'actor's point of view'. To use Schutz's terminology, they attempt to describe how social actors understand structures and constraints from within the 'natural attitude'. Indeed, one might argue from this perspective that the focus of housing researchers on power relations is unnecessarily limiting, and may even be scientifically wrong, if people do not see their own actions in these terms. One can contrast this with the claim made by Marx, which often has a residual presence through another variety of 'ontological gerrymandering' in poststructuralist writings, that the analyst has a more complete understanding of the hidden forces that shape individual action. It is difficult to avoid favouring one or other of these perspectives in conducting sociological research on what happens in housing agencies, the experiences of tenants in inner-city estates or how government policies are presented by the media.

Finally, it is worth making the point that, although the issues raised by philosophers like Schutz, Kuhn and Wittgenstein are both abstract and difficult, it can only benefit housing researchers to engage with these debates about constructionism. They not only encourage greater reflexivity and analytical rigour in conducting theoretically-informed research, but they also force us to think more clearly about how we design and conduct empirical research, and make us aware of a wider range of methods, and sociological literatures. Whatever philosophical or political position you adopt, the debates reviewed in this chapter are highly relevant to housing research.

Notes

1   Comte (1974) also argued that observations were meaningless unless understood in terms of a theory, so the differences between positivism and critical realism can be overstated.

2   For an overview of epistemological positions in social science, see Hughes and Sharrock (1997).

3   For examples, see Chapters 14 and 15 in Ritzer (1992). They include Anthony Giddens' structuration theory, and Pierre Bourdieu's writings on *habitus*.

4   One reason for the warm reception given to *The Social Construction of Reality* in the late 1960s is that it gives the impression that it is possible to change established institutions even though they exert massive coercive power over the individual. They note that 'because they are historical products of human activity, all socially constructed universes change, and the change is brought about by the concrete actions of human beings' (Berger and Luckman, 1967: 134).

5   It is ironic that Foucault is often presented as a conventionally radical thinker in support of a realist analysis of how society discriminates against marginal groups. In fact, his main objective as a thinker was to challenge and destabilise the claim of any group or theorist to have privileged access to the truth.

6   For an introductory text which advances Wittgensteinian arguments against scepticism using everyday examples, see Phillips (1996).

7   For an informative review of the sociology of scientific knowledge, see Shapin (1995).

8   For a discussion of this issue in relation to science, see Lynch (1996).

9   The key theoretical statement of labelling theory is Becker's *Outsiders* (1973; first published in 1963); the second edition contains a chapter responding to criticisms by structural

sociologists. Schneider (1985) provides a useful overview of social problems research with discussion of the empirical studies published by the journal *Social Problems*. For debates in this field, see the collections edited by Best (1989) and Holstein and Miller (1993).

10   For an extended argument in relation to the relativist position in the sociology of science, see Lynch (1993). This also takes issue with the way philosophers and sociologists associated with the 'strong programme' have used Wittgenstein to develop a relativist or sceptical argument against science. For a summary of this debate, see Pleasants (1997).

11   Pollner's argument was more complex in that he demonstrated how sociologists switched between two models in talking about deviance. The first was what Schutz would call the natural attitude; the second was a realist understanding of deviance that would exist even if it was not recognised.

12   It is tempting to conclude that how one understands the issue of construction in philosophical terms makes no difference to how sociologists collect or analyse data (an argument for keeping unnecessary philosophical debates out of sociology). On the other hand, it will be apparent that how one understands the issue of method cannot be separated from epistemological considerations. See, for example, the traditions reviewed in my text on qualitative research (2001).

13   For a review of 'postmodern ethnography', see Travers (2001), ch. 8.

References

Agger, B. (1993), 'The Problem with Social Problems: From Social Constructionism to Critical Theory', in J. Holstein and G. Miller (eds), *Reconsidering Social Constructionism: Debates in Social Problems Theory*, New York: Aldine de Gruyter: 281–300.

Becker, H. (1973), *Outsiders: Studies in the Sociology of Deviance* (2nd edn), New York: The Free Press.

Berger, P. and Luckmann, T. (1967), *The Social Construction of Reality*, London: Allen Lane.

Best, J. (1989), *Images of Issues: Typifying Contemporary Social Problems*, New York: Aldine de Gruyter.

Best, J. (1993), 'But Seriously Folks: The Limitations of the Strict Constructionist Interpretation of Social Problems', in J. Holstein and G. Miller (eds), *Reconsidering Social Constructionism: Debates in Social Problems Theory*, New York: Aldine de Gruyter: 129–50.

Bhaskar, R. (1978), *A Realist Theory of Science*, Brighton: Harvester Press.

Bloor, D. (1991), *Knowledge and Social Imagery*, Chicago: University of Chicago Press.

Bloor, D. and Barnes, B. (1982), 'Relativism, Rationalism and the Sociology of Scientific Knowledge', in M. Hollis and S. Lukes (eds), *Rationality and Relativism*, Oxford: Blackwell: 21–47.

Bogen, D. and Lynch, M. (1993), 'Do We Need a General Theory of Social Problems?', in J. Holstein and G. Miller (eds), *Reconsidering Social Constructionism: Debates in Social Problems Theory*, New York: Aldine de Gruyter: 213–40.

Comte, A. (1974 [1830]), *The Positive Philosophy*, New York: AMS Press.

Derrida, J. (1976), *Of Grammatology*, Baltimore: Johns Hopkins University Press.

Durkheim, E. (1966), *The Rules of Sociological Method*, New York: Free Press.

Foucault, M. (1972), *The Archaeology of Knowledge*, London: Tavistock.

Garfinkel, H. (1967), 'Passing and the Managed Achievement of Sex Status in an Inter-sexed Person', in H. Garfinkel (ed.), *Studies in Ethnomethodology*, New Jersey: Prentice Hall: 116–85.

Grathoff, R. (1978), *The Theory of Social Action: The Correspondence Between Alfred Schutz and Talcott Parsons*, Bloomington: Indiana University Press.

Halfpenny, P. (1982), *Positivism and Sociology*, London: Allen and Unwin.

Hastings, A. (1999), 'Discourse and Urban Change: Introduction to the Special Issue', *Urban Studies*, Vol. 36, No. 1 (special issue on discourse and urban change): 7–12

Holstein, J. and Miller, G. (1993), *Reconsidering Social Constructionism: Debates in Social Problems Theory*, New York: Aldine de Gruyter.

Hughes, J. and Sharrock, W. (1997), *The Philosophy of Social Research*, London: Longman.

Jacobs, K. and Manzi, T. (2000), 'Evaluating the social constructionist approach in housing research', *Housing, Theory and Society*, Vol. 17, No. 1: 35–42.

Kuhn, T. (1996 [1962]), *The Structure of Scientific Revolutions*, Chicago: University of Chicago Press.

Lazarsfeld, P. and Rosenberg, M. (eds) (1955), *The Language of Social Research*, New York: The Free Press.

Lynch, M. (1993), *Scientific Practice and Ordinary Action: Ethnomethodology and Social Studies of Science*, Cambridge: Cambridge University Press.

Lynch, M. (1996), 'Detoxifying the "Poison Pen Effect"', in A. Ross (ed.), *Science Wars*. Durham, NC: Duke University Press: 238–58.

Marlaire, C. and Maynard, D. (1993), 'Social Problems and the Organization of Talk and Interaction', in J. Holstein and G. Miller (eds), *Reconsidering Social Constructionism: Debates in Social Problems Theory*, New York: Aldine de Gruyter: 173–98.

Phillips, D.Z. (1996), *Introduction to Philosophy*, Oxford: Blackwell.

Pleasants, N. (1997), 'The Post-positivist Dispute in Social Studies of science and its Bearing on Social Theory', *Theory, Culture and Society*, Vol. 14: 143–55.

Pollner, M. (1974), 'Sociological and Common-sense Models of the Labelling Process', in R. Turner (ed.), *Ethnomethodology*, Harmondsworth: Penguin: 27–40.

Pollner, M. (1991), 'Left of Ethnomethodology', *American Sociological Review*, Vol. 56: 370–80.

Ritzer, G. (1992), *Sociological Theory* (3rd edn), New York: McGraw-Hill.

Sayer, A. (1984), *Method in Social Science: A Realist Approach*, London: Routledge.

Schneider, J. (1985), 'Social Problems: The Constructionist View', *Annual Review of Sociology*, Vol. 11: 209–29.

Schutz, A. (1967), *The Phenomenology of the Social World*, Evanston, IL: Northwestern University Press.

Schutz, A. (1973), *Collected Papers I: The Problem of Social Reality*, The Hague: Martinus Nijhoff.

Shapin, S. (1995), 'Here and everywhere: sociology of scientific knowledge', *Annual Review of Sociology*, Vol. 21: 289–321.

Travers, M. (2001), *Qualitative Research Through Case Studies*, London: Sage.

Weber, M. (1978), *Economy and Society*, Berkeley: University of California Press.

Weber, M. (1991), 'Politics as a Vocation', in H. Gerth and C. Wright Mills (eds), *From Max Weber: Essays in Sociology*, London: Routledge: 77–128.

Winch, P. (1977), 'Understanding a Primitive Society', in B. Wilson (ed.), *Rationality*, Oxford: Blackwell: 78–111.

Winch, P. (1988 [1958]), *The Idea of a Social Science and its Relation to Philosophy*, London: Routledge.
Wittgenstein, L. (1958), *Philosophical Investigations*, Oxford: Blackwell.
Woolgar, S. and Pawluch, D. (1985), 'Ontological Gerrymandering', *Social Problems*, Vol. 32: 214–27.

Chapter 3

Relativism, Subjectivity and the Self: A Critique of Social Constructionism

Peter King

I feel somewhat uncomfortable having been invited to criticise an idea and a research agenda within a book aimed at showcasing recent work in this area. It is as if I have been invited into someone's house and then asked to criticise their choice of furniture and fittings. My concern therefore is not to appear ungrateful to my hosts, but also to do as I am told.

This problem is compounded for me because there are some elements of social constructionism that I agree with, particularly the emphasis on subjectivity and meaning. I have written extensively on these areas (King, 1996, 1998, 2003, 2004) and I am therefore keen to see these areas developed further. The shift away from positivism and simple structural or agency explanations is an important and necessary shift in housing research, leaving behind the shallow ideology that housing relates to nothing more than production and consumption. I am not therefore implacably opposed to the aims of social constructionists. However, I do see it as being problematic and with serious limitations. I believe that we can appreciate subjectivity and meaning in dwelling environments more fruitfully without the adoption of the relativism and rather etiolated view of the self that is common to social constructionism.

My aim in this chapter is to concentrate on what I consider to be the main criticisms of social constructionism. The critique presented here is quite often general and not specifically related to the housing literature. The reason for this is the obvious one, that these general criticisms apply equally well to the housing literature. Indeed, on the few occasions when the housing literature has been criticised the emphasis has been on the problems of relativism and the connection between strong and weak versions of constructionism.

But there is a further reason for a more general criticism that goes beyond the housing literature. If the housing literature seeks to connect with the broader social constructionist corpus it has to respond to the general criticisms of the theory and thus how it remains appropriate as a method for housing researchers.

It seems to me that hitherto (and this book will undoubtedly go a long way to altering this perception) social constructionism has been used in a rather straightforward manner and housing researchers have not grappled with these criticisms. This may be because, with a few exceptions (Somerville, 2002; Somerville and Bengtsson, 2002a, 2002b) they have largely been unchallenged. It would be fair to say, then, that the criticisms presented here are rather conventional, and will not come as much of a surprise to anyone reasonably conversant with the literature. But they are serious criticisms and will be need to be countered if social constructionist research is to develop further.

What I wish to do in this chapter is, firstly, to consider the manner in which housing researchers have used social constructionism. I shall argue that much work has been undertaken on discourse analysis and language, even though there have been some attempts to consider other forms of social constructionism such as symbolic interactionism and the sociology of power. This being the main form in the housing literature, I shall offer some criticisms of this approach. I shall then look at what I consider to be the two main general criticisms of social construction. I begin with a critique of the coherence of relativism, and then move to the relationship between strong and weak constructionism. This is an important relationship, in that the weak version of social constructionism is seen by its proponents as offering a defence against the charge of relativism. It is also the version subscribed to by most housing researchers using the theory. This leads me into a brief discussion of the nature of social constructionism as a theory. Social constructionism is commonly criticised for its view of the self and subjectivity, and so accordingly I consider this issue. The next section considers what the appeal of social constructionism might be. I conclude with some comments about the usefulness of social constructionism and how certain concerns might be developed in a different manner. My aim throughout is to be, so to speak, constructive and to try and offer a rounded discussion rather than one that is entirely negative. Nevertheless my purpose is to demonstrate where social constructionism is at its weakest.

How Housing Researchers Use Social Constructionism

One of the aims of this book appears to be to suggest that the use of social constructionism can be broadened out beyond its current development in housing research. Kemeny (2002a) has suggested that social constructionism, despite what the recent literature would imply, is not merely discourse analysis. He admits that most researchers have taken social construction to mean this, and consequently the majority of articles published have focused on the application of discourse analysis to housing phenomena. But he argues that social constructionism can also encompass the construction of social problems, social interactionism and the sociology of power. Kemeny is clearly correct to state this overemphasis on discourse and the consequent neglect of symbolic interaction and other forms, and he points to his own work as examples of where power and hegemony as well as symbolic interaction is both discussed and actively applied (Kemeny, 1992, 1995, 2002b). There is a need therefore to broaden the theoretical remit to include other forms of social constructionism, and one can hope that this book will serve to demonstrate this.

My interest here, however, is to try to understand why discourse analysis might be so dominant. Perhaps the most obvious reason for the dominance of discourse analysis is the simple fact that nearly all definitions of social constructionism begin with the centrality of discourse for the construction of social relations. The editors in their introduction to this volume (Ch. 1: 3) define social construction as '*our access* to the material world is mediated through language and discourse'. Jacobs and Manzi (2000: 36) suggest that 'actors do not merely provide descriptions of events, but are themselves constitutive of wider discourses and conflicts'. This importance on discourse is also stressed by Clapham (2002), who, in his discussion on housing pathways, offers perhaps the fullest discussion of the importance of discourse and language to social constructionism. He states, 'Meaning is produced, reproduced, altered and transformed through language and discourse (Clapham, 2002: 61), and 'language and knowledge are not copies of reality, but constitute reality, each language constructing specific aspects of reality in its own way' (Clapham, 2002: 61). Even Kemeny himself, when defining his theoretical framework that stresses hegemony and power relations, suggests that hegemony is achieved 'in the social construction of negotiated orders in which modes of discourse and frames of debate are established' (Kemeny, 1992: 127). Therefore, even if we wish to stress the sociology of power we still return to discourse and language as the means by which this is to be achieved. We therefore ought not to be surprised of the preponderance of studies that emphasise discourse.

My second reason for the emphasis on discourse is more speculative, and quite simply relates to the fact that discourse, in both its narrow and broad senses, has been central to social theory for the last 25 years or so. The legacy of postmodernism, post-structuralism and deconstruction is a stressing of the centrality of discourse and language (Derrida, 1976; Foucault, 1970, 1972). Discourse, we might say, is quite simply the most fashionable concept to expound.

But this latter point does not mean it has no credibility: theories become fashionable because they have wide appeal, and this might be because they have some merit. What I wish to do then is to consider discourse briefly to show what its appeal might be. But I shall show that what appeals is precisely its biggest fault.

Discourse can be used in two ways: in the narrow sense as discourse *analysis*; and in the broader sense as discourse *theory*. Whilst I would argue that most researchers have explicitly used discourse analysis, when discourse is tied explicitly into social constructionism it taps into the broader theory.

Discourse analysis has been increasingly used by housing researchers and has produced some interesting and innovative studies (Franklin, 2001; Hastings, 2000; Jacobs and Manzi, 1996; Marston, 2002; Saugeres, 1999; Watt and Jacobs, 2000). This work, following Fairclough (1992), suggests that discourse can be seen as both textual analysis, in the sense of the structure of language, and as specific to particular situations. We can therefore identify specific housing discourses which might pertain to notions of quality (Franklin, 2001) or to housing management (Saugeres, 1999). These discourses, it is argued, can exclude or include individuals and groups depending on their facility with that language and their ability to manipulate it. By their very nature these studies are specific and relate to a given context. But they do connect with a broader theoretical sense of the role of discourse, and this is where the link with social constructionism becomes especially evident.

Discourse theory goes further than merely an analysis of language, and has been developed into a thorough critique of ideology, hegemony and power relations. As Torfing (1999) has suggested, discourse is now used in a wider sense than mere texts and language. Following Derrida (1978) and Laclau (1993), Torfing defines discourse as 'a decentred structure in which meaning is constantly negotiated and constructed' (Torfing, 1999: 40). The structure is seen as 'an ensemble of signifying sequences' (Torfing, 1999: 40) and allows for the inclusion of both physical and non-physical objects.

Torfing argues that discourse theory offers an anti-essentialist view of the individual subject as constantly constructed and reconstructed socially:

> In sharp contrast to the essentialist conception of identity, discourse analysis emphasises the construction of social identity in and through hegemonic practices of articulation, which partially fixes the meaning of social identities by inscribing them in the differential system of a certain discourse (Torfing, 1999: 41).

What this means is that the subject's identity is created by discourse, as much as discourse is created by individual subjects in speech acts. We might therefore suggest that a social housing discourse is not merely a label or a description of practices. Rather, in addition to these policies and practices, it is made up of the dwellings, the tenants, those organisations representing tenants, landlords, the surrounding environment, and the perceptions of the wider society. What is important to note here is that this discourse is not static, but is decentred, contingent and shifting. The discourse alters as the context changes, but the discourse also helps to change the context. An example of this is the manner in which the National Housing Federation (NHF), the lobby group for English housing associations, has sought to 'rebrand' its members' activities. In response to survey evidence that social housing was for 'losers' and that working in the sector was not seen to be of high status, the NHF employed branding consultants to try to remodel the image of housing associations as being 'iN [sic] business for neighbourhoods' (NHF, 2003). The clear aim here is to try to assert a new identity for social housing providers in the face of structural, political and ideological obstacles. Social housing was seen, rightly or wrongly, to be failing, so, it is suggested, its role should appear to change. The underlying assumption here is that there is no given role for housing associations, but that it has to respond to changing political and ideological conditions, and hence create a new discourse for itself.

This use of discourse appears to offer a powerful theoretical tool for considering analysing housing phenomena. The notion of housing discourse(s) can be used to consider power relations within and outside housing structures, and to assess these structures within the home. Indeed the idea of a housing discourse can enclose a huge range of activities and interests, including not just the structures of provision and consumption, but such issues as privacy, insularity, intimacy, childcare, property rights, gender, the use of technology, architectural styles, employment patterns, mobility, and so on. Housing discourse, we might suggest, can include many, and perhaps most, aspects of housing phenomena.

Now this might seem to be a key benefit of any theorisation – if a theory can enclose much that is significant, then it might be useful. Hence we could suggest the virtue of discourse is that it presents a theoretical tool to link microanalysis with broad structural developments and to provide researchers with a means of linking all this together.

But might we not also argue that a theory that excludes nothing – everything is part of discourse – actually offers us little room for analysis. As an example, we can compare discourse with theories of power. Most theories of power, be they Marxist, liberal or Foucauldian, are based around a dichotomy or dialectic of inclusion and exclusion. They involve the theorising of mechanisms for discriminating between individuals and groups: they are means by which we can state the relations between, these individuals and groups, *and seek to explain them.* However, it seems to me that with discourse all we can do is to list what it includes, or to state what we mean by the term, and then come to some generalities about its pervasiveness. What we are left with then is to say that 'discourse is everywhere and everything is discourse', which perhaps does not take us terribly far. As Somerville and Bengtsson (2002b) have suggested, a key problem with the manner in which discourse is defined and deployed is that it excludes nothing. It is reducible to a catch-all banality and consequently allows for no real theoretical development. I would argue, therefore, that the appeal of discourse analysis is also its greatest fault, namely, its inclusiveness: everything is discourse, but as a result nothing is differentiated from anything else. However, this does not offer a fatal blow to social constructionism; it merely states the limitations of one particular avenue of housing research. In order to offer a more fundamental challenge to social constructionism I wish to consider some more general arguments which challenge the very coherence of the theory.

The Problems of Relativism

The philosophical basis of social constructionism is the idea that social reality is not fixed and objectively determined, but is, rather, contingent and relative. The belief is that social reality is constructed through social interaction and is not pre-given. This assertion is a controversial one and if it is found to be lacking would seriously damage social constructionism. It is an issue that it is not possible to do full justice to in the limited space allowed. I shall therefore offer only a few observations on the coherence of relativism and what this implies for the subject under discussion.

Relativism is the belief that there are no necessary or objective relations in the world (Rorty, 1989). A less strong version would be to state that not all relations are necessarily objective (see the discussion on weak constructionism below). Whatever the version of relativism, however, all would suggest that our relations with the world are mediated and not objective in the sense of us seeing the world 'as it is', or even that there is a world 'as it is' for us to see. The world, so to speak, is created by our interactions with it and with those we share it with. As we shall see below, there is an active debate about the extent of relativism within social constructionism. But what I wish to do here is present a number of problems with relativism that apply whether one takes a strong or weak constructionist position.

Nozick (2001) has offered one of the most concerted attacks on relativism and social constructionism in recent years. He attacks both the coherence of relativism, as well as speculating on why social construction appeals (see below). His starting point is to show the essentially illogical incoherence of relativism. The belief that 'all social facts are relative' is, according to Nozick, problematical in that it relies on the fact that *at least one* statement is not relative, namely the statement 'all social facts are relative' itself. In order for the theory that all things are relative to hold, this one statement has to be universally true. But if it this is the case, what is its special status: relativists would need to explain what it is about this statement which means that it holds while all others do not? What this shows is that what we have to do here is to rely on non-contingency to justify the existence of the contingent. We have to be able to assert the veracity of the truth content of the sentence 'all social facts are relative' in order to state that all other social facts are contingent and arbitrary. But, why should we not see this statement as equally contingent? Or, indeed, if this statement is true, why not all others about the social world?

If we apply the same contingency criteria to social constructionism, cannot we then ask what status this theory has? Is it a social fact, and if so, why has it got a special status over and above those (apparently contingent) arguments that there are objective social facts? If the statement that 'all social relations are contingent' is true then it is incumbent on social constructionists to tell us why the statement 'all social facts are independent of their perception' is not also?

But to do this social constructionists are faced with a severe problem: to assert the special status of the claim that 'all social relations are contingent' *they cannot rely on non-contingency as an argument*. It is not permissible for them to rely on some further statement or meta-theory on which to ground their statement or theory. Social constructionists, like deconstructionists and other post-structuralists (Benton and Craib, 2001), cannot rely on any external mechanism on which to adjudicate their arguments. If we follow this argument through we have to concur with Nozick when he states that 'There cannot be social construction all the way down' (Nozick, 2001: 26). Ultimately a social construction has to rest on something:

> it cannot be that *everything* is a social construction; it cannot be that all truths and facts things are social constructions. For to say something, for instance gender differences are a social construction is to say that there are social

processes that cause this phenomena to exist or to take a certain form. If gender differences are a social construction, then it is a fact that certain existing social processes exist and produce those gender divisions. And *that* fact – that those social processes exist and produce gender divisions – is not itself a social construction. (Or if it too is a social construction, the fact that it is, is not itself a social construction.) (Nozick, 2001: 26).

The obvious way of dealing with Nozick's challenge is to agree with him and suggest that indeed social construction rests on certain things, and this is the very strategy taken by constructionists such as Jacobs and Manzi (2000). In effect they argue that instead of all facts being contingent, it is merely some facts.

But before I deal with this modified position I want to make two further points about relativism raised by Nozick. Firstly, I want to question what it is that social constructionism actually purports to do. What do we achieve by asserting the contingency of social relations? So, for example, if we suggest that homelessness is a contingent relation what have we then achieved? As Nozick states, just because a fact is relative does not make it alterable. We may suggest that homelessness is a social construction, but this does not make it go away, or in any manner alter the social relations in which a homeless household finds itself. We cannot simply close our eyes and say that it does not exist. Rather, the consequences of homelessness are as palpable and affecting whether we see it as contingent and arbitrary or not. Calling homelessness a social construct does not make it any easier to alter.

The last point I wish to make about relativism, again relying on Nozick's discussion, is: at what point can we say something is a social fact? If all Western developed nations have a recognisable state and a recognisable housing market, and if all states have some recognisable form of regulation of housing markets, cannot we not then state (without making any comments about any specific set of social arrangements) that these are social facts? Surely, we can argue, it is a matter of fact that certain social institutions that we call the welfare (or interventionist) state and the housing markets exist as real entities. These entities may take different forms in different communities, and do so for specific reasons, but they are recognisably of the same functional character. They therefore exist as social facts.

Weak or Strong?

I am aware that the argument above can be criticised for taking the extreme position on relativism and assuming that this is the one taken by housing researchers. However, as the editors state in their introduction, and as they have stated elsewhere (Jacobs and Manzi, 2000), most housing researchers do not deny the existence of an objective world. The position taken is rather to differentiate between 'ideas and concepts, which are socially constructed, and the social and spatial processes, which have a material existence' (Jacobs and Manzi, 2000: 38). Therefore the material world does not depend entirely on our perception of it. These social constructionists are not arguing against an objective world, they are merely stating that it is mediated by human thought and language. This being the case, it could be argued that Nozick's critique of relativism no longer applies. Housing researchers have rejected total relativism and instead have taken up a partial position in which only ideas and concepts are relative.

However, this does not really allow them to escape from the criticism of incoherence. In stating that it is only ideas and concepts that are relative, it would appear to follow that the statement 'ideas and concepts are socially constructed' is itself relative. What Jacobs and Manzi have explicitly done is to state that their conception of social construction is itself a construction, and is therefore open to question. What gives their statements particular status compared to all other ideas and concepts? If they are to assert that their concept is true they must therefore concur that some (if not all) ideas and concepts are objective. They would then need to explain why some ideas are true whilst others are contingent, or accept that all ideas are true and thus in effect reject constructionism altogether.

But, in any case, why should some things be socially constructed and not others? What is the relation between the constructed and non-constructed? Does one rely on the other, and, if this is the case, what impact does the constructed have on the non-constructed entities? As an example, what sense does it make to talk about housing markets without the concept of property rights? Many, if not all, material entities take on their particular significance because of the ideas and concepts we attach to them, and we cannot effectively separate concept from object in our perception of them. So, we might conclude, if we do not have unmediated access to the material world in what way does it make sense to talk about an objective world? And if so, where does this leave Jacobs and Manzi's distinction between strong and weak constructionism?

Putting this the other way round, if we start to move away from relativism and suggest that not all things are contingent, on what basis can we decide that some still are, and why should we seek to do so? What is so efficacious about this argument that we would wish to retain a part of it? If *everything* is not constructed why should *anything* be? What is so useful about social constructionism that we would want to maintain that some things are social constructs but others are not? One fears that the distinction is to try and maintain some semblance of intellectual coherence, rather than any worked out logically coherent position.

In order to maintain the weak position there needs to be some explanation of the relationship between the contingent and the absolute. Social constructionists would need to explain how the absolute retains its status when it comes into contact with the contingent and vice versa. If thought is contingent, how are we to separate this from the things themselves? If the whole essence of subjective or phenomenological approaches is to demonstrate the unfathomable interlinkage of objective/material and subjective/ideal, then how can we tenably assert that social constructionism is only partial *and retain a commitment to the social construction of reality*. Once we separate out the objective and subjective as distinct realms do not we throw away any of the insights of this approach?

A Total Explanation?

I would like now to turn briefly to a related contradiction in the social constructionist position. It appears to me that social construction, to be consistent, must offer a total explanation of social phenomena. It purports to explain the nature of social relations and offer a means of explaining them that will hold regardless of circumstance. We cannot argue that social constructionism applies to certain phenomena and not to others: we cannot say that (the idea or concept of) homelessness is a social construction but (the idea or concept of) housing markets are not (and maintain a consistent theory). Social construction appears therefore to be an absolute theory. But, as we have seen, the basis of social constructionism is to eschew absolute positions and replace them with contingency. We cannot therefore assert that social constructionism will always hold true without questioning the epistemological basis of the theory itself. In this sense social constructionism is like deconstruction (Lilla, 2002). It is a theory that argues against absolute and immutable positions. However, it cannot do so without calling into question its own basis. Just as much as deconstruction collapses when you try to deconstruct it, so social construction is presented with an insurmountable difficulty once we see it as a construction. Theoretically speaking, then, if it is a construction its premises cannot be said to hold with any certainty and this calls into question the very notion that anything else is a construct. Social construction must therefore be presented as a total explanation, but, as soon as it is, it contradicts itself.

A Real Self?

Another serious criticism of social constructionism is what it implies about the nature of individual human subjects. If the objective world is mediated through language and discourse (and the broader the definition of discourse the more problematical this becomes) what does this say about us as individuals? Is social constructionism not tantamount to the belief that we as individuals are constructed socially too?

Critical realists such as Archer (2000) criticise social constructionism for neglecting the embodied nature of selfhood. She argues that individual subjects are embodied and relate materially to their environment as well as linguistically. We have other senses apart from speech and these other senses are equally necessary for us to understand the world. Human bodies have properties and powers of their own that are active in an environment that is wider than the concept of 'society's conversation' that constructionists claim is the creator of a sense of self (Archer, 2000). She argues that social constructionists rely merely on a sense of passive selfhood, rather than the fully conscious *and* embodied self.

In order to relate socially we must have a fairly sophisticated transmitter/ receiver and the cognitive means to understand what is being said to us (as indeed do others in order to understand us). What this suggests is that we have some set of common equipment which allows us to make common and predicted responses. In order to relate to each other, human beings need to be sophisticated communication machines who are also able to process information in a manner compatible with others. For social relations to take place we must be active agents rather than passive bodies. This is not fully taken account of by social constructionism, which tends to stress the external construction of reality. We are asked to believe that reality is negotiated outside of the individual and within social relations, but without anything other than contingent relations for them to negotiate within. Where is the solidity, the framework; how would any person know they should get out of bed?

Social construction can be seen as an attempt to find the organising principle for intersubjectivity. It is discourse that becomes this organising principle. Yet this is really a chimera, in that discourse is both the input and the output. The world of the social constructionist is like that in the film *The Matrix* (1999, Wachowski Brothers) but without the machines to run it. *The Matrix* presents a completely socially constructed world where individuals respond to external stimuli, based on an extreme imbalance of power where machines and artificial intelligence are in control. It is, of course, an extremely unappealing vision of totally passive human beings plugged into an intersubjective neural network. But even here, we should note two things. The makers of *The Matrix*, even when they portray humans as being no longer born but grown and who spend their entire lives in vats attached to a computer network, still assume that individuals have substantial mental equipment in order to relate to the matrix they are plugged into. Take the humans out of their vats and they become competent beings. Secondly, the film shows that a socially constructed world requires an organising principle, in this case the machines. The difference between the film and social constructionism is that the latter argues that the organising principle is the intersubjectivity itself: the matrix rather than the machines. What purports to determine social relations is, in effect, those very same social relations. It is this very lack of a coordinating mechanism that leads its proponents to continually row back from their initial statements about the contingent nature of social relations which are created by discourse.

The problem, it seems to me, is that social constructionism explains selfhood and subjectivity as if it only derives from social relations, and hence individual subjects are empty vessels filled up through discourse. The problem here is essentially a confusion between subjectivity and *inter*subjectivity, between individual actions and the mechanisms through which individuals cooperate. But much of what is significant in subjective relations about housing is deeply private and personal. Moreover it is non-linguistic, relating rather to our emotions, passions and embodiment (Bachelard, 1969; King, 2004). Our relations with our dwelling might be seen as 'social' in the particular sense that it embodies shared values, traits and customs, but what we use our housing for is particular to us. Indeed, the subjectivity that is expressed in our housing is geared towards separating us from others, even when it is done in culturally common ways. What we use our dwelling for is to insulate us from the social and to explore our relations with others in ways that need no sanction beyond our own choices (King, 2004). Our housing allows us to distinguish ourselves, and this demonstrates a high degree of cognitive competence that is denied by social construction.

What is the Appeal of Social Construction?

The final issue I want to look at is to consider what is the appeal of social constructionism. I am aware that this is perhaps dangerous, as I risk impugning the motives of those who support the theory. It is also highly speculative to try and suggest why someone might be attracted to or seek to use a particular body of theory. However, my intention is not to impugn or cast any doubts on motives – what I suggest as motives are entirely reasonable in themselves. I believe, though, that my attempt to consider the appeal sheds some further light on the theory and leads to a further common criticism of social constructionism.

Nozick (2001) argues that the appeal of social constructionism arises from a desire to transform society. If we insist that certain things are arbitrary and contingent we then have grounds for changing them. Is it not the case, argues Nozick, that what we see as social constructions are those things we seek to change? Nozick suggests that referring to something as a social construction denudes that object of any authority, and thus it can (and therefore ought to) be changed. He suggests, and a trawl of the housing literature bears this out, that social constructions are never 'good things' or to be welcomed, but are things that the writer wishes to alter, change or improve. Thus, to call something a social construction gives one the grounds to challenge it.

But what basis is there for deciding that certain things are contingent and others not? There is not an issue if we suggest all things are constructions, but we have already seen the problems involved in this, and that most housing researchers do not wish to sustain this view. I would argue that, assuming consistency of argument, social constructionists have no grounds for determining which entities are constructions and which are not. This is for the reasons given above, that to state something so categorical necessitates foundations and these are precisely what is foresworn in both the strong and weak versions of social construction. The problem is that social construction cannot allow itself an external means of adjudicating on its outcomes without negating itself.

This leads us to the common accusation that is levelled against relativist positions, namely, that they are, in some way, reactionary. Critics such as Finkielkraut (2000) and Habermas (1987) comment on the effects of relativism and contingency in denying any sense of universal human values such as natural rights. Similarly, Lash (2002: 110) is critical of the 'incessant construction' he sees at the heart of philosophies of postmodernism. He suggests that this 'incessant construction' does not allow for the 'we' of solidarity to develop. The emphasis on flux, on making and remaking our social relations, militates against any togetherness and sense of commonality.

Lash suggests we need something that is 'given', something that lies behind our actions, determining what we perceive and therefore how we construct our lives and negotiate with others. Lash too appears to be saying, like Nozick, that constructions cannot go the whole way down, that at some point we need to fix our relations to some concrete point. From this point we can create solidarity and solidity.

Social constructionists would and do contest this accusation. For example, Manzi (2002) contends that social construction can come to an understanding of the nature of oppression by looking at its different forms and definitions. They can therefore come to an understanding of the nature and effects of oppression.

But why should they wish to do anything about it? On what basis could they argue that oppression is wrong? To return to the discussion on weak constructionism, what is the basis for objecting to the idea (and therefore construct) that 'oppression is always wrong?' Those of us who adhere to notions of human rights or the inviolability of the human person can offer a fairly straightforward answer to this – oppression violates a persons rights and we have a responsibility to ensure this does not occur (King, 2003) – but this position is denied the social constructionist who is forced to argue on a case by case basis that a particular social relation is wrong or to be contested, presumably on the basis of its particular effects. However, I would suggest that it is more likely that they would wish to deal with oppression because they object to it on other grounds, and it is found to be a social construct precisely because of this objection. But this is only possible if there is some foundation or ground by which to judge social relations, and this foundation must necessarily be outside the particular social relations themselves. Again it appears that social constructionists must continually row back from the implications of their theory, and seek to place themselves again on solid ground.

Conclusions

I began this chapter by claiming I am not an enemy of social construction, and have then spent the rest of my allotted space pointing out its faults. What I wish to do by way of a conclusion, then, is to point to those elements within social constructionism that might be developed, albeit perhaps in ways different from that preferred by its proponents.

The first thing to state is that I would not wish, despite what I consider strong arguments against it, to see researchers stop using social constructionism.

Housing research is lacking enough in theoretical and conceptual analysis without me putting anyone off. The majority of theoretical work in housing has been of a social constructionist bent, and what I hope this book engenders is a healthier environment for theoretical discussion, and one where there are a number of different schools or positions. So I welcome the focus that this book should give, but also seek to encourage opposing views to it.

To my mind the main benefits that have derived from social constructionist housing research have been the emphasis it has placed on subjectivity and the role that individual subjects play in housing structures. This has been a necessary corrective to the political economy and structural approaches that dominated until the mid-1990s, but which still have some influence even now. What has always worried me is the tendency to see housing as a matter of production and consumption at the macro level (King, 1996) and thus to neglect what happens, so to speak, behind the front door. Once dwellings are built they are then used, and this aspect has always seemed to me to be more interesting. Social constructionists have begun to develop a better appreciation of how housing is used and what individual subjects do in it. A particular example is Clapham (2002, and this volume) with his housing pathways approach. This offers a much more sophisticated and nuanced view of how individual households change as a result of their own choices and because of changing political, social and economic factors.

But in my view the discussion on subjectivity has not gone far enough. As I have argued above, when social constructionists discuss subjectivity they really mean intersubjectivity, and not the meanings, use and emotional attachments that individuals place on their own dwelling. What I believe is called for, then, is a return to the original phenomenological concerns that led to the development of social constructionism. This would involve a return to the original sources of phenomenology such as Heidegger (1962) and Husserl (1970), as well as to other writers such as Bachelard (1969) and Merleau-Ponty (1962). This would lead to a greater concern with the meaning that we attach to our housing and to an appreciation of use, difference and similarity. One of the fascinating aspects of housing from a conceptual point of view is that it is both ubiquitous and unique: that we all do it, and in similar ways, but we all do it differently and seek to separate ourselves from others (King, 2004). Social constructionism has gone some way to appreciating this, but it cannot go much further because of its predominant concern for the mechanisms of intersubjectivity.

What social constructionism has done, therefore, is to open up the field of housing research and demonstrate the potential for theoretical analysis that links housing into a broader body of theory. It shows that there is potential for us to learn from other disciplines and fields, but also that we can talk to other theoreticians about issues of common interest. Social construction has many faults, and to my mind these are fatal to it as research project, but it does point a way forward, to theoretically informed housing research linked to an established social theory. I wish it well, if only so I have something to argue against.

References

Archer, M. (2000), *Being Human: The Problem of Agency*, Cambridge: Cambridge University Press.

Bachelard, G. (1969), *The Poetics of Space*, Boston: Beacon.

Benton, T. and Craib, I. (2001), *Philosophy of Social Science: The Philosophical Foundations of Social Thought*, Basingstoke: Palgrave.

Clapham, D. (2002), 'Housing Pathways: A Post Modern Analytical Approach', *Housing, Theory and Society*, Vol. 19, No. 2: 57–68.

Derrida, J. (1976), *Of Grammatology*, Baltimore: Johns Hopkins University Press.

Derrida, J. (1978), *Writing and Difference*, London: Routledge.

Fairclough, N. (1992), *Discourse and Social Change*, Cambridge: Polity Press.

Finkielkraut, A. (2000), *In the Name of Humanity: Reflections on the Twentieth Century*, New York: Columbia University Press.

Foucault, M. (1970), *The Order of Things: An Archaeology of the Human Sciences*, London, Routledge.

Foucault, M. (1972), *The Archaeology of Knowledge*, London: Routledge.

Franklin, B. (2001), 'Discourses of Design: Perspectives on the Meaning of Housing Quality and "Good" Housing Design', *Housing, Theory and Society*, Vol. 18, Nos 1–2: 79–92.

Habermas, J. (1987), *The Philosophical Discourse of Modernity*, Oxford: Blackwell.

Hastings, A. (2000), 'Discourse Analysis: What Does it Offer Housing Studies?', *Housing, Theory and Society*, Vol. 17, No. 3: 131–9.

Heidegger, M. (1962), *Being and Time*, Oxford: Blackwell.

Husserl, E. (1970), *The Crisis of European Science and Transcendental Phenomenology: An Introduction to Phenomenology*, Evanston, IL: Northwestern University Press.

Jacobs, K. and Manzi, T. (1996), 'Discourse and Policy Change: The Significance of Language for Housing Research', *Housing Studies*, Vol. 11, No. 4: 543–60.

Jacobs, K. and Manzi, T. (2000), 'Evaluating the Social Constructionist Paradigm in Housing Research', *Housing, Theory and Society*, Vol. 17, No. 1: 35–42.

Kemeny, J. (1992), *Housing and Social Theory*, London: Routledge.

Kemeny, J. (1995), *From Public Housing to Social Renting: Rental Policy Strategies in Comparative Perspective*, London: Routledge.

Kemeny, J. (2002a), 'Re-inventing the Wheel? The Interactional Basis for Social Constructionism', *Housing, Theory and Society*, Vol. 19, Nos 3–4: 140–41.

Kemeny, J. (2002b), 'Society versus the State', *Housing, Theory and Society*, Vol. 19, Nos 3–4: 185–95.

King, P. (1996), *The Limits of Housing Policy: A Philosophical Investigation*, London: Middlesex University Press.

King, P. (1998), *Housing, Individuals and the State: The Morality of Government Intervention*, London: Routledge.

King, P. (2003), *A Social Philosophy of Housing*, Aldershot: Ashgate.

King, P. (2004), *Private Dwelling: Speculation of the Use of Housing*, London: Routledge.

Laclau, E. (1993), 'Discourse', in R. Goodin P. and Pettit (eds), *A Companion to Contemporary Political Philosophy*, Oxford: Blackwell: 431–7.

Lash, S. (2002), *Critique of Information*, London: Sage.

Lilla, M. (2002), *Reckless Minds: Intellectuals in Politics*, New York: New York Review of Books.

Manzi, T. (2002), 'Construction, Realism and Housing Theory', *Housing, Theory and Society*, Vol. 19, Nos 3–4: 144–5.

Marston, G. (2002), 'Critical Discourse Analysis and Policy-orientated Housing Research', *Housing, Theory and Society*, Vol. 19, No. 2: 82–91.

Merleau-Ponty, M. (1962), *Phenomenology of Perception*, London: Routledge.

National Housing Federation (2003), *In Business for Neighbourhoods: Action for Change*, London: National Housing Federation.

Nozick, R. (2001), *Invariance: The Structure of the Objective World*, Cambridge, MA: Belknap/ Harvard University Press.

Rorty, R. (1989), *Contingency, Irony and Solidarity*, Cambridge: Cambridge University Press.

Saugeres, L. (1999), 'The Social Construction of Housing Management Discourse: Objectivity, Rationality and Everyday Practice', *Housing, Theory and Society*, Vol. 16, No. 2: 93–105.

Somerville, P. (2002), 'But Why Social Constructionism?', *Housing, Theory and Society*, Vol. 19, No. 2: 78–9.

Somerville, P. and Bengtsson, B. (2002a), 'Constructionism, Realism and Housing Theory', *Housing, Theory and Society*, Vol. 19, Nos 3–4: 121–36.

Somerville, P. and Bengtsson, B. (2002b), 'Understanding contextualised rational action – authors' response', *Housing, Theory and Society*, Vol. 19, Nos 3–4: 148–52.

Torfing, J. (1999), *New Theories of Discourse: Laclau, Mouffe and Žižek*, Oxford: Blackwell.

Watt, P. and Jacobs, K. (2000), 'Discourses of Social Exclusion', *Housing, Theory and Society*, Vol. 17, No. 1: 14–26.

Chapter 4

Extending Constructionist Social Problems to the Study of Housing Problems[1]

Jim Kemeny

Introduction

Constructionist approaches to social problems emerged in the 1970s as an alternative to traditional positivist approaches. This took place in the context of the hermeneutic turn in sociology that began in the late 1950s as a reaction against the dominance of positivism and structuralism, in which the individual was treated as a cultural dope, merely holding values and playing out roles dictated by structural imperatives. It included critiques of quantitative methods, including large-scale interview surveys, that involved a turn to qualitative methods and the study of ongoing social situations.

In the 1970s constructionist explanations of social problems became well established in sociology as well as in a number of related disciplines, notably social psychology, social work and, to a lesser extent, political science. Early constructionist approaches to social problems were strongly influenced by symbolic interactionism and particularly by labelling theory, and only later, in response to the linguistic turn, became increasingly influenced by discourse analysis. Constructionist approaches eventually became established as the dominant approach to the study of social problems. This did not take place without considerable and often heated debate, both internally but also from the proponents of competing perspectives, notably from Marxists and feminists, and later also from postmodernists, as well as, of course, from traditional positivists.

Yet perversely, the debates raging around how to study social problems passed housing studies by, almost entirely unnoticed. This is particularly puzzling since housing studies has always been – and remains to this day – a very applied field in which the study of housing problems and the housing policies to deal with them comprise a central concern. Until very recently housing research has remained a bastion of traditional positivism, heavily influenced by structuralist explanations and strongly oriented toward the use of quantitative methods, including the use of official statistics and the collection of additional data through large-scale interview surveys.

In stark contrast to constructionist social problems, constructionist housing problems, developing in the late 1990s, have been barely influenced at all by the symbolic interactionist origins of constructionist social problems. Instead they have been dominated by discourse analysis from the very start, something for which they have been, with justification, criticised. Research on constructionist housing problems therefore has much to learn from interactionism in general but in particular from the successes and failures of constructionist social problems analysis.

Both Travers (Chapter 2) and King (Chapter 3) examine the constructionist approach to social problems from philosophical perspectives. In this chapter I review the development of constructionist approaches to social problems in sociology and consider what lessons this holds for housing researchers in terms of the development of the still embryonic constructionist approach to housing problems. I show how each new perspective that has arisen – including Marxist and feminist perspectives on social problems – has been strongly influenced and formed by perspectives and approaches that were dominant at the time of its initial development. I conclude with a discussion of the rapidly growing – but still modest – corpus of constructionist housing problems research.

The State of Social Problems Research in the Pre-constructionist Era

Early post-war research on social problems was, as might be expected, framed by positivist debates that did not question the taken-for-granted 'naturalness' of the hierarchy of social problems in a society at any one point in time. This is not to say that there was no awareness of the disjuncture between actual 'real' social problems, such as unemployment or crime, and people's perceptions of the relative importance of different social problems.

Thus, R.K. Merton, one of the early post-war pioneers of the study of social problems, distinguished between 'manifest' social problems, those objective conditions which are recognised as social problems and 'latent' social problems, those that are at odds with dominant values and therefore not recognised as social problems (Merton, 1951; see also Merton, 1971). Merton thereby remained firmly in the functionalist tradition of neither problematising whether a condition was or was not a 'real' social problem nor being interested in its genesis or explaining processes underlying the emergence of social problems.

Another pioneer of social problems analysis was C. Wright Mills. His book *The Sociological Imagination* opens by making a distinction between 'the personal troubles of milieu' and 'the public issues of social structure' (Mills, 1959: 14). One of Mills' key concerns in relating personal troubles and public issues was the ways in which the former became translated into the latter. For Mills the sociological imagination comprised in large part the ability to show how personal milieux translated into large-scale social structures and societal issues, or as he put it:

> *Issues* have to do with matters that transcend these local environments of the individual and the range of his inner life. They have to do with the organization of many such milieux into the institutions of a historical society as a whole, with the ways in which various milieux overlap and interpenetrate to form the larger structure of social and historical life (Mills, 1959: 15).

Yet his analysis did not significantly progress Merton's distinction between manifest and latent social problems. He remained in the positivist tradition of arguing that there were objective social issues caused by the proliferation of personal troubles on such a large scale as to create institutional scale issues. Thus, individuals may become unemployed or divorced and these are personal troubles. But when unemployment becomes widespread or when the divorce rate in the first four years of marriage is 250 out of every 1,000 then, Mills decrees, personal troubles become a social issue. This remains today the dominant perspective on housing problems such as homelessness, overcrowding, housing and ill health and segregation.

Mills saw the definition of social issues out of personal troubles as a key area of conflict and disagreement in society, a view not developed significantly further until interactionist approaches to the construction of social problems were developed in the early 1970s. It did, however, foreshadow the construction of social problems literature in its focus on power, a focus that permeated Mills' own brand of sociological analysis.

The 1960s comprised something of a transition decade between positivist and constructionist approaches to social problems, a transition well captured in Horton's (1966) analytical discussion of conflict versus consensus approaches to the study of social problems. Howard Becker, who was to emerge as one of the strongest proponents of labelling theory and interactionism, exemplifies this transitional period with a focus on value-conflict that never the less retained an insistence that social problems have an objective existence (Becker, 1966).

One of the earliest analyses of a particular social problem that was both firmly in the conflict perspective on social problems as well as clearly influenced by early constructionist ideas was Gusfield's (1963) study of the US temperance movement and its impact on prohibition politics. Working at Chicago University, Gusfield acknowledged his indebtedness to Herbert Blumer, Everett Hughes and Erving Goffman, as well as citing the work of Mead, Lemert, Matza, Garfinkel, Stone and others who either had made, or were going to make, major contributions to symbolic interactionism, ethnomethodology and constructionism. He was also influenced by Murray Edelman, the leading political scientist in the field of symbolic politics. The theory Gusfield presented in his last chapter to explain temperance politics in the US was significantly entitled 'a dramatistic theory of status politics', indicating the influence on his thinking at Chicago of Goffman's 'dramaturgical sociology' and Edelman's 'symbolic politics'.

Symbolic Crusade, with its strong emphasis on the rise and fall of the influence of the temperance movement on alcohol policy, was extensively cited in the early constructionist literature on social problems. It marked a significant shift in the empirical analysis of social problems, towards a focus on social movements, interest groups, lobbying and rhetoric. The publication of *Symbolic Crusade* came at a time when the social sciences were ready for a shift from structural functionalism to conflict perspectives and in particular to constructionism: a shift that was reflected in a variety of complementary lines of development and that together represented nothing less than a sea change in the direction of the development of the social sciences (though not in housing research). Deviance and labelling theory (Becker, 1963; Matza, 1964), interpersonal negotiation (Roth, 1962; Scheff, 1968), symbolic interactionism (Blumer, 1962, 1969), negotiated order theory (Strauss et al., 1963), ethnomethodology (Garfinkel, 1967), and of course constructionism (Berger and Luckmann, 1966) all contributed to this.

The Emergence of the Constructionist Approach to Social Problems

The large corpus of creative research from the symbolic interactionist inspired constructionist perspective that emerged during the 1960s – both conceptual/ methodological and in a number of different applied fields such as deviance, medical sociology, mental illness, professions, and organisational studies – paved the way for the development of the generic application that came to be known as constructionist social problems. As late as 1971 the first edition of what in later editions became a highly influential reader in constructionist social problems (Rubington and Weinberg, 1971) provided an overview of five perspectives on social problems, one of which was based on labelling theory. A sea change was already becoming apparent.

By the early 1970s the Society for the Study of Social Problems had a number of constructionists in senior positions, including its president, Edwin Lemert, and members of the executive committee such as Helena Znaniecka Lopata and Arlene Daniels, as well as on its editorial and publications committee, notably Earl Rubington, Severyn Bruyn and John Kitsuse. The journal *Social Problems* was to become the prime vehicle for the promotion of the perspective. The formative contributions to constructionist social problems, all published in the journal *Social Problems*, were Blumer (1971), Kitsuse and Spector (1973) and Spector and Kitsuse (1973).

The first clear statement was the defining contribution of Blumer in 'Social Problems as Collective Behavior' (1971). Blumer was already a major founding figure in symbolic interactionism, indeed having coined the term (Blumer, 1962, 1969). In this paper he formulates the main issues in what was to become the constructionist approach to the study of social problems: that social theory is unable to detect what is or is not a social problem, but rather follows societal definitions; that societal definitions and not objective conditions define what becomes identified as a social problem; and that to understand why some social problems become seen as important it is necessary to study the social processes of definition formation: as Blumer puts it: 'A social problem is always a focal point for the operation of divergent and conflicting interests, intentions, and objectives' (Blumer, 1971: 301).

Blumer then proceeds to outline the study of social problems as *the study of a social process* – their emergence, legitimation, mobilization of support, the formation of an official plan of action, and the transformation of the official plan in its empirical implementation. This article, published over 30 years ago, remains to this day an accurate preliminary statement of the constructionist social problems perspective.

The next development was the work of Kitsuse and Spector (1973) and Spector and Kitsuse (1973), comprising, in effect, a combined two-part article. Kitsuse and Spector (1973) review the work of Merton as the exemplar of a functionalist approach and the much earlier Fuller and Myers (1941) as exemplifying value-conflict approaches – both Mills (1959) and Becker (1966) being more recent examples.

Taking their starting point in the above analysis, Spector and Kitsuse (1973) then argue that social problems are better understood in terms of process rather than as static conditions. That is, social problems have a career, or what Fuller and Myers (1941) termed a 'natural history'. They define social problems as 'the activities of groups making assertions of grievances and claims to organisations, agencies, and institutions about some putative conditions' (Spector and Kitsuse, 1973: 146). They argue that each claims-making activity has a different natural history that can be analysed, described and its contingencies delineated.

One of the consequences of this was to shift the focus towards the way social problems were formulated, the discourse that was used to describe and present problems and how solutions in terms of policy measures were constructed and sometimes implemented. Much of this had been present in embryonic form in the Gusfield (1963) study of the temperance movement, albeit without conceptual explication. However, with the impetus provided by Spector and Kitsuse (1973) the 1970s and 1980s were something of a boom period in constructionist social problems, generating a remarkably wide range of studies of social problems from the constructionist perspective (see, for example, Best, 1989; Gusfield, 1981; Holstein and Miller, 1989; Manis, 1976; Schneider and Kitsuse, 1984; Schneider, 1985a; Useem and Zald, 1982). This extensive literature covered a remarkably wide range of social problems. Besides dealing with more established social problems such as alcohol (Gusfield, 1963) and child abuse (Pfol, 1977) a wide range of putative, or supposed, conditions were taken up, ranging from margarine as a social problem (Ball and Lilly, 1982) to nuclear power (Uzeem and Zald, 1982).

The literature sometimes gave the impression that social problems remain stable: what changes is the way they are seen, defined and successfully lobbied over by pressure groups, and it is this which determines how much of a social problem they are seen as, and how much urgent attention they are given by policy makers – and during which historical periods. This approach often implied – and sometimes explicitly argued – that problems pass through 'rise and fall' cycles (Ball and Lilly, 1982; Hilgartner and Bosk, 1988; Jacobs, Kemeny and Manzi, 1999; see also Adam, 1975 for an example from feminism, also discussed below).

Critiques of Constructionist Social Problems

1   The 'Ontological Gerrymandering' Debate

A major internal, constructionist, critique of the approach that had come to be taken by many constructionist social problems analysts was that by Woolgar and Pawluch (1985a). They argued that much constructionist social problems work defines the particular social problem being studied as fixed and immutable: as a natural and unvarying phenomenon. In doing this they disregard the problematic nature of the condition itself. Conditions cease to be *putative* in the manner argued by Spector and Kitsuse (1973) but rather become transformed into 'real' conditions. The social problem doesn't change, only the way it is defined and understood changes, with policy towards the social problem changing as a result. Constructionist social problems is therefore *selectively constructionist* – repressing discussion of any changes in the problem or putative condition itself while centre-staging the flux and instability of definitions that arise out of conflicting interest groups as they struggle to gain ascendancy in policy-making towards the fixed condition. Woolgar and Pawluch (1985a) characterise this selectivity as 'ontological gerrymandering':

> Viewed as a practical accomplishment, both theoretical statements and empirical case studies manipulate a boundary, making certain phenomena problematic while leaving others unproblematic. We call the main strategy for managing this boundary ontological gerrymandering (Woolgar and Pawluch,1985a: 214).

At one level this is an impossible critique to answer. Everything clearly cannot be problematised at the same time in the same study: 'bracketing' is and will always be an essential strategic decision for researchers to make. All research must by definition make 'certain phenomena problematic while leaving others unproblematic' and to call 'managing this boundary ontological gerrymandering' is a gross and unacceptable distortion.

In spite of this, the critique did strike a cord and engendered a lively and valuable debate (Pfol, 1985; Schneider, 1985b; Hazelrigg, 1985; Woolgar and Pawluch, 1985b). This centred on the extent to which constructionism needs to be more reflexively aware of its own constructions as the first step towards understanding how to accomplish these practices in what came to be known as constructionist studies of science (see, for example, Latour and Woolgar, 1979).

Largely as a response to this debate, Ibarra and Kitsuse (1993) proposed a revision of the original Spector and Kitsuse (1973) formulation. They argued that Spector and Kitsuse's use of the expression *putative* (or supposed) *condition* was unfortunate as it allowed constructionist analyses to do just that – accept the condition as 'supposed' and by default treat it as an immutable background 'fact' against which claims and related discourses were constructed. The result is to risk the analyst 'going native' and rather than 'bracketing' claims accepting one or another pressure group's vernacular construct – their definition of the situation – as the condition. As they put it:

> When constructionist theorizing blurs the basic distinction between vernacular resources and analytic constructs, it invites an indiscriminate fusion of mundane and theoretical perspectives that, among other things, leads to a retreat from the distinctive task of description posed by Spector and Kitsuse's formulation (Ibarra and Kitsuse, 1993: 33–4).

Ibarra and Kitsuse propose that, to avoid these problems, the study of social problems should be understood as the study of what they term *claims-making activities*. That is, 'claims-making constitutes social problems' (Ibarra and Kitsuse, 1993: 34).

This is a helpful clarification in so far as the idea of a putative condition is replaced by the varying depictions of a social problem through the claims-making of interests and pressure groups: a claims-making that the constructionist analysing the process of claims-making has to bracket. The study of social problems thereby becomes the study of claims-making and the ability of pressure groups to impose their conflicting definitions of reality on public discourse and ultimately on the policy making process. Social problems thereby become the *accomplishments* of successful claims-making. One consequence of this reformulation is that the study of social problems begins to look very much like the study of social movements (see Mauss, 1975 for a conflict perspective that predates constructionist social problems).

2   Constructionist Critiques of Ibarra and Kitsuse

Ibarra and Kitsuse's paper (1993) was published in a reader on social problems (Holstein and Miller, 1993a) which included a number of useful articles on constructionist social problems. Some of these were 'sympathetic critiques' from constructionists who were critical of one aspect or another of Ibarra and Kitsuse's thesis. Others were more generally critical of constructionist social problems as an approach to the study of social problems. In this and the following section my argument draws heavily on both of these types of contribution to that reader.

Having established the need to focus on the accomplishment of social problems rather than adopt essentialist definitions of social problems, Ibarra and Kitsuse (1993) go on to limit the usefulness of their approach by overly focusing on rhetoric and discourse at the expense of the *accomplishment* of social problems and the construction of policy outcomes that the deemed politically-appropriate 'treatment' results in. They focus on the deployment of rhetorical idioms such as 'lifestyle', 'choice', 'tolerance', 'empowerment' and various 'isms' (sexism, racism, ageism, etc.), and the employment of rhetorical and counter-rhetorical strategies, motifs, and claims-making styles ('scientific', 'political', etc.). Finally, they discuss three social settings in which rhetoric is deployed which they label *media, legal-political* and *academic.* They thereby give the misleading impression that the study of social problems is mainly – or even entirely – the study of rhetoric and discourse, rather than the outcome of this in terms of legislation, institutional changes and other practical results of successful claims-making that accomplish social problems.

Ibarra and Kitsuse have clearly been heavily influenced by the linguistic turn and the almost obsessive focus on discourses. Accomplishing a social problem is more than just generating sound bytes that drown out alternative definitions, or as Gubrium's (1993: 96) critique of Ibarra and Kitsuse in the Holstein and Miller reader puts it 'mere words in media'. It is more than simply establishing a hegemonic definition in the media and in government rhetoric of the social problem, then deciding what should be done about it and how it should be 'treated'. The accomplishment lies in the mundane interactions that produce the soundbites – 'the hard reality work that generates the publicity, belying the "moves" and "gambits" that media audiences hear or read about' (Gubrium, 1993: 96).

Equally important, I would argue, is that mundane interactions are essential to devising and implementing ways of dealing with an accomplished social problem. This includes the successful construction of institutions for dealing with a social problem in ways that are congruent with the hegemonic definition – in terms of administrative measures, new or revised legislation, the creation of appropriate institutions (workhouse, hospital, orphanage, etc.), and the creation and nurturing of pressure groups to effectively monitor policy and sustain pressure on policy making (Jacobs, Kemeny and Manzi, 2003a).

Another problem with Ibarra and Kitsuse's approach is the generalisation and abstraction of the category 'members', together with the implication that the public are passive receptors of discourse or, as Gubrium (1993: 97) puts it, that 'the public, in general or in particular, [are] simply a nebulous mass of receptive agents'. To address this, it is necessary to conduct studies into the social worlds of particularly situated individuals, and it is here that interactionist *in situ* studies of non-problems – such as gender or ethnic discrimination – come into their own. Gubrium's own study of the caregivers of

Alzheimer sufferers suggests that many are sceptical and critical of professional pronouncements in an area of care where they as caregivers can lay claim to considerable collective experience and expertise (Gubrium, 1987).

A related problem is what Schneider (1993) sees as a too rigid separation of constructionists' own analyses from those of the members they study – a separation into 'them', the unwitting objects of study, and 'us', the scientists who explain 'their' behaviour to each other, to the subjects they study and to the wider and generalised public. He argues that constructionists cannot leave their own assumptions and memberships of various social groups as unproblematic and unproblematised. I would add that doing so is symptomatic of research strategies that have too little involvement in the life-worlds being researched.

This leads naturally on to the argument of Best (1993), the last of the constructionist critiques of Ibarra and Kitsuse. He argues that Kitsuse has always taken what Best terms a 'strong' constructionist approach, even though many interpretations of Kitsuse's work prefer to adopt a 'weak' constructionist interpretation, and the Ibarra and Kitsuse thesis is an example of strong constructionist analysis. Here housing researchers will at last be on familiar ground (see Somerville and Bengtsson, 2002). Best argues that the Ibarra and Kitsuse injunction to only study claims-making and not make any assumptions about reality is unrealistic: he elegantly demonstrates this with reference to research on Satanism, where the study of claims of Satanist conspiracy is difficult if not impossible to conduct, simply because it is often impossible to identify any such Satanist groups. But all this might mean is that either the claim is false or that the conspiracies are very successful in remaining secret. Best argues, therefore, that weak constructionism is preferable since 'It will not, to be sure, meet the strict constructionists' tests for epistemological consistency, but it just might help us understand how social problems emerge and develop' (Best, 1993: 144).

3   *External Critiques of the Constructionist Social Problems Perspective*

Apart from objectivist critiques that social problems are in fact objective conditions, three broad streams of 'external' critiques can be identified: postmodern, Marxist and feminist (for a useful overview see Miller, 1993). Interestingly, each of these critiques is made from quite different epistemological positions.[2]

(a) Postmodern critiques Postmodern critiques share the constructionist social problems approach in focusing on narratives but deny the validity of trying to

develop systematic macrosocial explanations of the sort constructionist social problems uses. Instead, they emphasise great variation that cannot be reduced to standardised pressure group positions that act and interact, as it were, on a policy chessboard. These variations may be local, psychological, or simply reflect a tendency to provide accounts that suit particular circumstances and contexts without any attempt to maintain consistency (for a discussion of postmodern critiques see Miller, 1993: 270–73).

Postmodernism can best be described as *deconstructionist*. That is, while it agrees with constructionist social problems that the world is socially constructed, it argues that the kinds of overarching societal constructions that constructionist social problems attempts to erect are too systemic. Instead, construction is less coordinated, more chaotic, local, and fragmented. Postmodernism therefore radically critiques constructionist social problems for its macrosocial focus, its overgeneralisation and its failure to grasp the complexity and micro-nature of the process of construction.

(b) Some Marxist and feminist critiques   Marxist and feminist critiques, by contrast, argue that constructionist social problems is itself too fragmenting, breaking down social problems into partial and episodic events unrelated to one another, whether it be acid rain, child abuse or AIDS (Agger, 1993). Both Marxist and feminist critiques seek less fragmentation and more conceptual and theoretical integration. They are also critical of the interactionist injunction to 'take the actor's definition'. They argue that capitalism and patriarchy both operate through hegemony, determining through the encouragement of false consciousness what come to be accepted as the important social problems of the time. So the very fragmentation that postmodernist critiques of constructionist social problems see as essential to understanding social problems is seen by both Marxists and feminists as the operation of hegemony to divert concern from the oppression of capitalism and patriarchy onto trivial social problems. In other respects, however, Marxism and feminism handle constructionism very differently.

(i)   Feminist critiques
Much feminist scholarship has been profoundly formed by phenomeno-logical constructionism. One of the key arguments is that sex is an ambiguous biological category and that gender is not a natural or biologically determined bipolar category. Rather it is multidimensional and socially constructed in many forms (Butler, 1990, 1993). Similarly, Adams (1975) analysis of the way 'women's place' in Britain swung back and forth

several times between being in the home (pre-1914, 1930s, 1950s) and being in wage-labour (1914–18, 1939–45, post-1950). This discussion is immediately recognisable as a constructionist analysis of social problems focusing on the cyclical 'rise and fall' thesis of female wage-labour.

Echoing a Goffman-inspired dramaturgical sociology, gender is to be understood as being 'performed' to be realised (or as interactionists might put it 'doing gender') and in the constructionist spirit has to be analysed as 'a process of engenderment'. There is therefore some justification for describing feminist critiques of constructionist social problems as internal constructionist critiques, albeit of a less than sympathetic kind than those reviewed above.

One of these is by Gordon (1993), whose critique arises from a strong constructionist perspective, accusing Ibarra and Kitsuse's thesis of being a kind of 'neo-objectivism'. By this Gordon means that for Ibarra and Kitsuse a social problem only exists in as much as there are voiced definitions coordinated and articulated by pressure groups and vested interests. Gordon argues that constructionism 'has allowed feminists to make visible how we and things are engendered' given that visibility is a complex phenomenon 'entirely bound up with the available and dominant technologies of representation and the social and 'personal' forms invisibility takes' (Gordon, 1993: 312).

Gordon argues that the main difference between feminist and internal constructionist critiques of constructionist social problems is that 'The feminist project is driven by critique and a desire to transform the social relationships of power that construct and deconstruct the world and its inhabitants in particular ways' (Gordon, 1993: 312). Here, then, we are back to some of the shared ground between Marxism and feminism.

(ii) Marxist critiques

Marxist critiques, in contrast to feminist critiques, have never been based on constructionist principles. Marxist responses to constructionism came early, and were therefore contemporaneous with constructionist social problems because as 'the new criminology' they were in considerable part counter-critical responses to labelling theory and other phenomenological and constructionist critiques of traditional criminology that had developed in the 1960s. Taylor et al. (1973) devotes much space to critiques of the leading societal-reaction theorists – Becker, Lemert, Matza, Cicourel and others. Taylor et al. (1973 – including the foreword by Gouldner) is still today the most detailed and incisive critique of this school.

The new criminology critiqued the older, conservative (positivist) approaches that labelling theory had also criticised. But in addition it criticised the more recent 'liberal' phenomenological approaches to deviance for being too social psychological and ignoring the way the structural features of capitalism formed and determined deviance (Taylor et al., 1974; Manders, 1975). So while acknowledging that labelling theory exposed the class basis of deviance it was dismissed as 'voyeurism' or as 'exposé criminology'. Foreshadowing the ontological gerrymandering debate, they argued – using the phenomenological argument of 'bracketing' – that constructionist social policy 'bracketed society and social theory' and, in so doing, diverted attention from structural Marxist explanations of deviance (Taylor et al., 1973, 1974). Nevertheless, the new criminology was very aware of the class basis on which capitalism rested and so was in part formed by labelling theory.

(iii) The new criminology and the new urban sociology

There are interesting similarities and differences with the housing critiques of the new urban sociology. The early social and political economy contributions (Pahl, 1969; Clarke and Ginsburg, 1975; Community Development Project, 1976; Edwards et al., 1976) were firmly anchored in a micro-class perspective on housing that examined the role of estate agents, financiers, brokers, solicitors etc. The original formulation of what came to be known as the 'urban managerial thesis' by Pahl (1969) argued for attention to be focused on those holding key mediating positions in housing and urban planning: building society managers, local government officers, planners, etc. However, unlike the new criminology, the 1970s new urban sociology did not have a corpus of constructionist or phenomenological housing research to mirror or critique since none then existed. If it had existed the urban managerial thesis might have found more support.

The turning point for the new urban sociology appears to have come at the conference on 'Urban Change and Conflict' held at the University of York in 1975, where 'urban managerialism' was one of the central foci (Harloe, 1975). The work of French Marxist structuralist urban sociology, notably Castells (1977, originally published in French in 1972) gained momentum at the expense of micro-based political economy class analysis. But Pahl came under heavy and sustained criticism, particularly from Marxists (Castells and Preteceille were both at the conference) for giving too much weight to agency and not enough to underlying structural factors (for an example of such a debate see the proceedings of the 1975 conference on 'Urban Change and Conflict' in Harloe, 1975). Pahl subsequently accepted this correction (Pahl, 1977), effectively consigning the urban managerial thesis to oblivion.

The problem facing those espousing the social basis of power approach to housing was that processes within housing institutions, and most especially the central and local state, were by definition secondary and derivative in nature and lacked an interactionist or constructionist tradition in housing and urban research to give it sufficient conceptual underpinning. A strong strand of societal reification therefore permeated analysis and inhibited a closer examination of the workings of social institutions, and in particular, decision-making processes in political institutions.

As a result, the new urban sociology generally, and the political economy of housing in particular, did not develop a well-defined class-based Marxist analysis but instead became closely identified with structural variants of Marxism in which class and agency rarely figured as an important explanatory factor. The result was to produce a rather ponderous and deterministic – and ultimately barren – structuralism (for critiques of 'class-purged' structural Marxism in urban and housing research see Kemeny, 1982 and 1984a).

Toward Developing Constructionist Housing Problems

We have seen how constructionist social problems began with strong phenomenological and symbolic interactionist impetus and only later – particularly since Ibarra and Kitsuse (1993) – was there a shift to discourse analysis as part of the linguistic turn in the social sciences. This can be even more clearly seen in the recent interest in 'collective action frames' or 'organizational framing' following Snow (1986). Although borrowing Goffman's (1974) concept of *Frame Analysis*, its application to social problems has been very much in the discourse analysis tradition, focusing on the ways in which social movements frame their aims and goals (Croteau and Hicks, 2003; Reese and Newcombe, 2003) and thereby become more effective at recruiting members and influencing policy making.

By contrast and as already noted, until very recently research on housing problems has been strongly positivist: unreflexively being drawn into the acceptance of government- and media-determined definitions of what comprise the main housing problems and how they should be 'solved'. Such research has been heavily dependent on the use of traditional large-scale interview surveys of the sort favoured by grant-giving government departments. Despite decades of constructionist social problems development and debate, until the late 1990s the influence of constructionist social problems had been all but nonexistent on research on housing problems. What we have had for the most part have been opportunistic studies of current political concerns, whether this be homelessness, negative equity, residential immobility, segregation, immigrant housing, residualisation, or a host of other politically-defined and policy-driven 'problems'.

Today, 30 years after the first flowering of the constructionist social problems perspective, now that at last constructionist research is being carried out on housing problems its trajectory has so far been the opposite to that of constructionist social problems. That is, it began its career with post-Foulcauldian discourse analysis and only recently has begun to show signs of being influenced by symbolic interactionism.[3]

The theoretical awakening of housing research which began with the new urban sociology and has exploded in the last decade or so has so far quite simply missed the influence of constructionist social problems based on the study of interpersonal – and indeed *intra*personal (reflective, self-examining) – interaction, a grounded theory approach and the use of *in situ* contextualised data collection and analysis.

The result is a very particular kind of constructionist housing problems research, based on the discourse analysis of policy documents and focused on discourse analyses of the housing policy spin of policy makers, the media, police, courts, lobbyists, etc. The early influence of discourse analysis on housing policy included such work as that by de Neufville and Barton (1987), indicating the yeoman imagery behind policies in the USA to favour owner occupation, Carlen (1994) on the maintenance of youth homelessness, and Sahlin's work on problem tenants (1996), social housing exclusion strategies (1995) and homelessness (1997).

Since then there has been a rapid expansion, notably the work of Hastings (1999), Hunter and Nixon (1999), Jacobs, Kemeny and Manzi (1999, 2003a, 2003b) and Arapoglou (2004). This approach has notable similarities to Ibarra and Kitsuse's restatement of constructionist social problems in discourse terms, a restatement that, as we have seen, came in for heavy criticism from symbolic interactionists and phenomenologists.

This has laid constructionist housing research open to the sort of critique that Somerville and Bengtsson (2002) have made of constructionism being merely about discourse, lacking a focus on agency and therefore being too socially deterministic. There is some irony in the fact that, without Somerville and Bengtsson realising it, their preferred alternative – derived from political science – of 'contextualised rational action' bears a remarkable resemblance to studies based on interactionist principles of individual agency and *in situ* studies of interaction in context.

The confusion of a discourse-based constructionist housing problems with the very different interactionist-based one reflects the continuing *lacuna* of the latter in housing research. That this is so comes as no surprise, given what we know of the differences between feminist-constructionist and Marxist critiques of constructionist social problems, not to mention the way the new urban sociology came to be dominated by a structural determinism that both neglected the individual level – albeit after considerable internal debate – and which was remarkably lacking in class-based analyses.

At the same time awareness is rapidly growing of the need to focus more on how housing policies are constructed in terms of the institutional arrangements that are put in place to 'deal with' particular definitions of what the housing problem 'really is'. The focus by Jacobs, Kemeny and Manzi (2003b) on the construction of a particular kind of policy implementation apparatus specifically adapted to certain kinds of problem definitions and policy solutions and of corresponding 'institutional practices' is in this sense an important marker along the way.

Housing research is in need of rectifying this imbalance by adding a range of *in situ* studies of housing problems to the already significant corpus of studies of discourses. This will involve going beyond research based exclusively on the analysis of document texts or the transcripts of qualitative interviews. Such methods, while of fundamental importance to discourse analysis, need to be understood as complementary to both direct and participant observation of interactional situations in which researchers take the actors' definitions and observe how social problems are defined by various interest groups and how policy is created and implemented on the ground.

It is important to bear in mind that the study of social situations as integral and dynamic wholes has always been a central tenet of interactionism. Following an epistemology known as 'grounded theory' (Glaser and Strauss, 1967), theory is inducted from a social situation, or a 'life-world' that the analyst enters, becomes immersed in and thoroughly imbued with, taking the actors' definitions and learning what it is to be a member.

This is sometimes referred to as *in situ* research or 'doing' whatever situated activity of everyday life is being researched ('doing everyday life' as Dietz et al., 1994 call it). The classic empirical research in this long and enduring tradition is precisely such participatory qualitative studies, of, for example, gangs (Thrasher, 1927; White, 1955), medical students (Becker et al., 1963), hotel communities (Prus and Irini, 1980), and other key studies too numerous to cite (for an extensive review and discussion see Prus, 1996). Methodology is strongly influenced by ethnology being qualitative, based on participant observation, in-depth interviews, and other methods of thorough familiarisation.

Much useful work can be done on studies of 'doing' housing policy making and housing policy implementation, and understanding these as *processes,* as achievements of interaction. A precursor of what such research could look like is Henderson and Karn's (1987) study of the allocation process in an English Midlands housing department. Using intensive observation techniques supplemented by interviewing and despite not being anchored in interactionist ethnology, this research provided important insights into the policy-implementation process around just how allocations are decided. Conducting research by being where the decisions are made and observing the processes of interaction that lead to the decisions being studied is a powerful methodology that has been neglected in housing research. Such a methodology becomes even more effective if is based on interactionist and constructionist principles, including taking actors' definitions and with an awareness of the work-practices and routines in the organisation as well as paying attention to the existence and stability or instability of negotiated orders.

Such studies can be done in a wide range of organisational settings: housing lobbies and pressure groups, government departments, media news determination, courts, police practices, etc. They can also be done at many levels of policy-making and implementation. Equally important they can be used to point up *processes*, such as labelling, developing new routines or area stigmatisation. A recent example of this is Roschelle and Kaufman (2004) based on a four-year ethnographic study of the inclusion and exclusion strategies of young homeless people.

Another aspect that a constructionist housing problems can fruitfully address is the study of consciousness-raising of the sort that feminists call for, notably around issues of problem invisibility and face-to-face processes of identity confirmation and definitional struggle. Research on what has come to be known as 'non-decision-making' in political science (Bachrach and Baratz, 1962; Crenson, 1971; Lukes, 1974) can be drawn on in this context.

Constructionist housing problems is in its infancy and poised for rapid expansion. It is crucial that we take this opportunity to learn the lessons from the earlier development of constructionist social problems so we benefit from the accumulated experience of a rich tradition while not repeating early mistakes. If we succeed, the 30-year 'constructionist slumber' of housing research will not have be wasted.

Notes

1   I would like to thank Jan Trost and other members of the Uppsala University IBF/Jane Seminar for helpful comments on an earlier version of this chapter.
2   Quite apart from the fact that each critique is not homogeneous but includes many different perspectives, which, of course, cannot be done justice in the limited space in this chapter.
3   An early exception was Kemeny's (1984b) study of the social construction of housing facts, based on Latour and Woolgar's (1979) classic symbolic interactionist study of the Salk Institute and how scientists in their everyday work negotiated how to interpret technical data so as to define what comprised 'fact' and what comprised irrelevant 'noise'.

References

Adams, R. (1975), *A Woman's Place: 1910–1975*, London: Chatto and Windus.

Agger, B. (1993), 'The Problem with Social Problems: From Social Constructionism to Critical Theory', in J. Holstein and G. Miller (eds), *Reconsidering Social Constructionism: Debates in Social Problems Theory*, New York: Aldine de Gruyter: 281–99.

Arapoglou, V. (2004), 'The Governance of Homelessness in Greece: Discourse and Power in the Study of Philanthropic Networks', *Critical Social Policy*, Vol. 24, No. 1: 102–26.

Bachrach, P. and Baratz, M. (1962), 'The Two Faces of Power', *American Political Science Review*, Vol. 56: 947–52.

Ball, R. and Lilly, R. (1982), 'The Menace of Margarine: The Rise and Fall of a Social Problem', *Social Problems*, Vol. 29, No. 5: 488–98.

Becker, H. (1963), *Outsiders: Studies in the Sociology of Deviance*, New York: Free Press.

Becker, H. (1966), *Social Problems: a Modern Approach*, New York: John Wiley.

Becker, H., Hughes, E. and Strauss, A. (1963), *Boys in White: Student Culture in Medical School*, Chicago: Chicago University Press.

Berger, P. and Luckmann, T. (1966), *The Social Construction of Reality: A Treatise in the Sociology of Knowledge*, New York: Doubleday.

Best, J. (ed.) (1989), *Images of Issues: Typifying Contemporary Social Problems*, New York: Aldine de Gruyter.

Best, J. (1993), 'But Seriously Folks: The Limitations of the Strict Constructionist Interpretation of Social Problems', in J. Holstein and G. Miller (eds), *Reconsidering Social Constructionism: Debates in Social Problems Theory*, New York: Aldine de Gruyter: 129–47.

Best, J. (ed.) (1994), *Troubling Children: Studies of Children and Social Problems*, New York: Aldine de Gruyter.

Blumer, H. (1962 [1937]), 'Society as Symbolic Interaction', in A.M. Rose (ed.), *Human Behaviour and Social Processes*, London: Routledge and Kegan Paul: 179–92.

Blumer, H. (1969), *Symbolic Interactionism: Perspective and Method*, Englewood Cliffs, NJ: Prentice Hall.

Blumer, H. (1971 [1937]), 'Social Problems as Collective Behavior', *Social Problems*, Vol. 18: 298–306.

Butler, J. (1990), *Gender Trouble: Feminism and the Subversion of Identity*, New York: Routledge.

Butler, J. (1993), *Bodies That Matter: On the Discursive Limits of 'Sex'*, New York: Routledge.

Carlen, P. (1994), 'The Governance of Homelessness: Legality, Lore and Lexicon in the Agency – Maintenance of Youth Homelessness', *Critical Social Policy*, Vol. 14, No. 2: 18–35.

Castells, M. (1977), *The Urban Question: a Marxist Approach*, London: Matthew Arnold.

Clarke, S. and Ginsburg, N. (1975), *Political Economy and the Housing Question: Papers Presented at the Housing Workshop of the Conference of Socialist Economics*, London: Political Economy of Housing Workshop.

Community Development Project (1976), *Profits Against Houses: an Alternative Guide to Housing Finance*, London: CDP Information and Intelligence Unit.

Crenson, M. (1971), *The Unpolitics of Air Pollution: A Study of Non-decision Making in the Cities*, Baltimore: Johns Hopkins University Press.

Croteau, D. and Hicks, L. (2003), 'Coalition Framing and the Challenge of a Consonant Frame Pyramid: The Case of a Collaborative Response to Homelessness', *Social Problems*, Vol. 50, No. 2: 251–72.

de Neufville, J. and Barton, S. (1987), 'Myths and the Definition of Policy Problems', *Policy Sciences*, Vol. 20: 181–201.

Dietz, M., Prus, R. and Shaffir, W. (1994), *Doing Everyday Life: Ethnography as Human Lived Experience*, Toronto, ON: Copp Clark Longman (Addison-Wesley).

Edwards, M., Gray, F., Merrett, S. and Swann, J. (1976), *Housing and Class in Britain: a Second Volume of Papers Presented at the Political Economy of Housing Workshop of the Conference of Socialist Economists*, London: Political Economy of Housing Workshop.

Fuller, R. and Myers R. (1941), 'Some Aspects of a Theory of Social Problems', *American Sociological Review*, Vol. 6: 24–32.

Garfinkel, H. (1967), *Studies in Ethnomethodology*, Englewood Cliffs, NJ: Prentice Hall.

Glaser, B. and Strauss, A. (1967), *The Discovery of Grounded Theory: Strategies for Qualitative Research*, Chicago: Aldine.

Goffman, E. (1974), *Frame Analysis: an Essay on the Organization of Experience*, Cambridge, MA: Harvard University Press.

Gordon, A. (1993), 'Twenty-two Theses on Social Constructionism: A Feminist Response to Ibarra and Kitsuse's "Proposal for the Study of Social Problems"', in J. Holstein and G. Miller (eds), *Reconsidering Social Constructionism: Debates in Social Problems Theory*, New York: Aldine de Gruyter: 301–26.

Gubrium, J. (1987), 'Structuring and Destructuring the Course of Illness: The Alzheimer's Disease Experience', *Sociology of Health and Illness*, Vol. 3, No. 1: 1–24.

Gubrium, J. (1993), 'For a Cautious Naturalism', in J. Holstein and G. Miller (eds), *Reconsidering Social Constructionism: Debates in Social Problems Theory*, New York: Aldine de Gruyter: 89–101.

Gusfield, J. (1963), *Symbolic Crusade: Status Politics and the American Temperance Movement*, Urbana: University of Illinois Press.

Gusfield, J. (1981), *The Culture of Public Problems: Drinking-driving and the Symbolic Order*, Chicago: University of Chicago Press.

Harloe, M. (ed.) (1975), *Proceedings of the Conference on Urban Change and Conflict*, London: Centre for Environmental Studies, Conference Paper 14.

Hastings, A. (1998), 'Connecting Linguistic Structures and Social Practices: A Discursive Approach to Social Policy Analysis', *Journal of Social Policy*, Vol. 27: 191–211.

Hazelrigg, L. (1985), 'Were it Not for Words', *Social Problems*, Vol. 32, No. 3: 234–7.

Henderson, J. and Karn, V. (1987), *Race, Class and State Housing*, Aldershot: Gower.

Hilgartner, S. and Bosk, C. (1988), 'The Rise and Fall of Social Problems: A Public Arenas Model', *American Journal of Sociology*, Vol. 94: 53–78.

Holstein, J. and Miller, G. (eds) (1989), *Perspectives on Social Problems*, Volume 1, Greenwich CT: JAI.

Holstein, J. and Miller, G. (eds) (1993a), *Reconsidering Social Constructionism: Debates in Social Problems Theory*, New York: Aldine de Gruyter.

Holstein, J. and Miller, G. (1993b), 'Reconsidering Social Constructionism', in J. Holstein and G. Miller (eds), *Reconsidering Social Constructionism: Debates in Social Problems Theory*, New York: Aldine de Gruyter: 5–23.

Horton, J. (1966), 'Order and Conflict Theories of Social Problems as Competing Ideologies', *American Journal of Sociology*, Vol. 71, No. 6: 701–13.

Hunter, C. and Nixon, J. (1999), 'The Discourse of Housing Debt: The Social Construction of Landlords, Lenders, Borrowers and Tenants', *Housing, Theory and Society*, Vol. 16, No. 4: 165–78.

Ibarra, P. and Kitsuse, J. (1993), 'Vernacular Constituents of Moral Discourse: An Interactionist Proposal for the Study of Social Problems', in J. Holstein and G. Miller (eds), *Reconsidering Social Constructionism: Debates in Social Problems Theory*, New York: Aldine de Gruyter: 25–58.

Jacobs, K., Kemeny, J. and Manzi, T. (1999), 'The Struggle to Define Homelessness: A Constructivist Approach', in D. Clapham and S. Hutson (eds), *Homelessness: Public Policies and Private Troubles*, London: Cassell: 11–28.

Jacobs, K., Kemeny, J. and Manzi, T. (2003a), 'Power, Discursive Space and Institutional Practices in the Construction of Housing Problems', *Housing Studies*, Vol. 18, No. 4: 429–46.

Jacobs, K., Kemeny, J. and Manzi, T. (2003b), 'Privileged or Exploited Council Tenants? The Discursive Change in Conservative Housing Policy from 1972–1980', *Policy and Politics*, Vol. 31, No. 3: 307–20.

Kemeny, J. (1982), 'A Critique and Reformulation of the New Urban Sociology', *Acta Sociologica*, Vol. 25, No. 4: 419–30.

Kemeny, J. (1984a), 'Economism in the New Urban Sociology: A Critique of Mullins' "Theoretical Perspectives on Australian Urbanisation"', *Australian and New Zealand Journal of Sociology*, Vol. 19, No. 3: 517–27.

Kemeny, J. (1984b), 'The Social Construction of Housing Facts', *Scandinavian Housing and Planning Research*, Vol. 1, No. 3: 149–64.

Kitsuse, J. and Spector, M. (1973), 'Toward a Sociology of Social Problems: Social Conditions, Value Judgements and Social Problems', *Social Problems*, Vol. 20, No. 4: 407–19.

Latour, B. and Woolgar, S. (1979), *Laboratory Life: the Social Construction of Scientific Facts*, New York: Sage.

Lukes, S. (1974), *Power: a Radical View*, London: Macmillan.

Manders, D. (1975), 'Labelling Theory and Social Reality: A Marxist Critique', *The Insurgent Sociologist*, Vol. 6, No. 1: 53–66.

Manis, J. (1976), *Analysing Social Problems*, New York: Praeger.

Mankoff, M. (ed.) (1972), *The Poverty of Progress: the Political Economy of American Social Problems*, New York: Holt, Rinehart and Winston.

Mason, G. (1976), 'You Have to Have Been There: The Method of Naturalistic Inquiry', in D. Thorns (ed.), *New Directions in Sociology*, Totowa, NJ: Rowman & Littlefield: 103–14.

Matza, D. (1964), *Delinquency and Drift*, New York: Wiley.

Mauss, A. (1975), *Social Problems as Social Movements*, Philadelphia: Lippincott.

Merton, R. (1951), *Social Theory and Social Structure*, Glencoe: Free Press.

Merton, R. (1971), 'Epilogue: Social Problems and Sociological Theory', in R.K. Merton and R. Nisbet (eds), *Contemporary Social Problems*, New York: Harcourt, Brace, Jovanovich: 793–864.

Miller, G. (1993), 'New Challenges to Social Constructionism: Alternative Perspectives on Social Problems Theory', in J. Holstein and G. Miller (eds), *Reconsidering Social Constructionism: Debates in Social Problems Theory*, New York: Aldine de Gruyter: 253–78.

Mills, C. Wright (1959), *The Sociological Imagination*, New York: Oxford University Press.

Pahl, R. (1969), *Whose City?*, Harmondsworth: Penguin.

Pahl, R. (1977), 'Managers, Technical Experts and the State: Forms of Mediation and Dominance in Urban and Regional Development', in M. Harloe (ed.), *Captive Cities: Studies in the Political Economy of Cities and Regions*, New York: Wiley.

Pfol, S. (1977), 'The "Discovery" of Child Abuse', *Social Problems*, Vol. 25: 310–24.

Pfol, S. (1985), 'Towards a Sociological Deconstruction of Social Problems', *Social Problems*, Vol. 32, No. 3: 228–32.

Prus, R. (1996), *Symbolic Interaction and Ethnographic Research: Intersubjectivity and the Study of the Human Lived Experience*, Albany: State University of New York Press.

Prus, R. and Styllianoss, I. (1980), *Hookers, Rounders, and Desk Clerks: The Social Organization of the Hotel Community*, Toronto, ON: Gage (reprinted by Sheffield Press, Salem, Wisconsin, 1988).

Reese, E. and Garnett N. (2003), 'Income Rights, Mothers' Rights or Workers' Rights?: Collective Action Frames, Organizational Ideologies and the American Welfare Rights Movement', *Social Problems*, Vol. 50, No. 2: 294–318.

Roschelle, A. and Kaufman, P. (2004), 'Fitting In and Fighting Back: Stigma Management Strategies among Homeless Kids', *Symbolic Interaction*, Vol. 27, No. 1: 23–46.

Roth, J. (1962), 'The Treatment of Tuberculosis as a Bargaining Process', in A. Rose (ed.), *Human Behavior and Social Processes: An Interactionist Approach*, London: Routledge.

Rubington, E. and Weinburg, M. (eds) (1971), *The Study of Social Problems: Five Perspectives*, New York: Oxford University Press.

Rubington, E. and Weinburg, M. (eds) (1977), *The Study of Social Problems: Five Perspectives*, 2nd edn, New York: Oxford University Press.

Sahlin, I. (1995), 'Strategies for Exclusion from Social Housing', *Housing Studies*, Vol. 11, No. 3: 381–401.

Sahlin, I. (1996), 'Discourses on Housing Problems and Problem Tenants', in R. Camastra and J. Smith (eds), *Housing: Levels of Perspective*, Amsterdam: AME (Amsterdam Study Centre for the Metropolitan Environment): 114–23.

Sahlin, I. (1997), 'Discipline and Border Control: Strategies for Tenant Control and Housing Exclusion', in M. Huth and T. Wright (eds), *International Critical Perspectives on Homelessness*, Westport, CT: Praeger Publishers: 139–52.

Scheff, T. (1968), 'Negotiating Reality: Notes on Power in the Assessment of Responsibility', *Social Problems*, Vol. 16, No. 1: 3–17.

Schneider, J. and Kitsuse. J. (eds) (1984), *Studies in the Sociology of Social Problems*, New Jersey: Ablex.

Schneider, J. (1985a), 'Social Problems Theory – The Constructionist View', in R. Turner (ed.), *Annual Review of Sociology*, Palo Alto, CA: Annual Reviews: 209–29.

Schneider, J. (1985b), 'Defining the Definitional Perspective on Social Problems', *Social Problems*, Vol. 32, No. 3: 232–4.

Schneider, J. (1993), '"Members Only": Reading the Constructionist Text', in J. Holstein and G. Miller (eds), *Reconsidering Social Constructionism: Debates in Social Problems Theory*, New York: Aldine de Gruyter: 103–14.

Snow, D.A., Burke Rochford Jr, E., Worden, S. and Benford, R. (1986), 'Frame Alignment Processes: Micromobilization and Movement Participation', *American Sociological Review*, Vol. 51: 464–81.

Somerville, P. and Bengtsson, B. (2002), 'Constructionism, Realism and Housing Theory' (including commentaries and author response), *Housing, Theory and Society*, Vol. 19, No. 3: 121–52.

Spector, M. and Kitsuse, J. (1973), 'Social Problems: A Reformulation', *Social Problems*, Vol. 21, No. 2: 145–59.

Spector, M. and Kitsuse, J. (1977), *Constructing Social Problems*, Hawthorne, NY: Aldine de Gruyter.

Strauss, A., Schatzman, L., Ehrlich, E., Bucher, R. and Sabshin, M. (1963), 'The Hospital and its Negotiated Order', in E. Friedson (ed.), *The Hospital in Modern Society*, New York: Free Press: 147–69.

Taylor, I. and Young, J. (1973), *The New Criminology For a Social Theory of Deviance*, London: Routledge and Kegan Paul.

Taylor, I., Walton, P. and Young, J. (1974), 'Advances Towards a Critical Criminology', *Theory and Society*, Vol. 1, No. 4: 441–76.

Thrasher, F. (1927), *The Gang*, Chicago: University of Chicago Press.

Useem, B. and Zald, M. (1982), 'From Pressure Group to Social Movement: Organizational Dilemmas of the Effort to Promote Nuclear Power', *Social Problems*, Vol. 30, No. 2: 144–56.

White, W. (1955), *Street Corner Society*, Chicago: University of Chicago Press.

Woolgar, S. and Pawluch, D. (1985a), 'Ontological Gerrymandering: The Anatomy of Social Problems Explanations', *Social Problems*, Vol. 32, No. 3: 214–27.

Woolgar, S. and Pawluch, D.(1985b), 'How Shall We Move Beyond Constructivism', *Social Problems*, Vol. 33, No. 2: 159–62.

Chapter 5

Constructing the Meaning of Social Exclusion as a Policy Metaphor

Greg Marston

Introduction

The role of housing in society is constructed in contradictory ways. Housing can be represented as a form of investment or as a fundamental human right, as a place of security, as promoting sustainability and as a very visible marker of status and wealth. These tensions are ever present in policy deliberations. On one level, housing policy is a contested public discourse involving political choices and administrative decisions about the appropriate division between public and private responsibility for the provision of adequate housing. There are regular debates in the Australian housing policy community, for example, about the appropriate mix of tax incentives, individual demand side subsidies and supply side measures in the housing system. At the heart of these policy debates are competing conceptions about the proper boundaries between social categories, such as the state, markets, citizens and civil society in the distribution of social and economic resources. Implicit in the discursive politics of housing policy are competing conceptions about what constitutes an ideal housing system. Every government programme presupposes an end of this kind – a type of person, community, organisation, society, or even world that is to be achieved (Dean, 1999: 33). As such, identifying the means and ends of housing provision remains an important task for the housing researcher and the policy analyst alike.

Competing conceptions about what constitutes a 'housing good' can be readily observed, however, housing research has traditionally failed to interrogate the assumptions and identity representations that constitute such debates (Bacchi, 1999; Marston, 2000; Hastings, 1998; Jacobs and Manzi, 1996). The legacy of positivism has meant that housing problems have been accepted by housing researchers as mirroring a pre-given reality. The primary task of the housing researcher working within this framework is understood as one of developing rigorous measures of housing affordability, for example, or identifying ways to improve the 'evidence-base' of housing policy (Marston and Watts, 2003). Surveys and statistics can and should inform us about the scope and scale of housing poverty. However, this form of policy knowledge tells us little about the value of these social goods or whether the unreflexive categories of being 'poor', 'disadvantaged', 'excluded' or 'dependent' have any real purchase on the accumulated experience and meanings of those that remain the silent objects of these studies.

In sum, a narrow range of research methods and an over-reliance on the discipline of economics has limited the depth and breadth of legitimate questions that housing researchers have explored. This situation is beginning to change. Recent attention to discourse and the politics of policy making reflects a recognition that the multidisciplinary field of housing research needs to engage with a more encompassing set of questions. Over the past decade, there has been growing interest among housing researchers in problematising the discursive framing of policy problems. Social constructionism and the methods of inquiry it gives rise to encourage researchers to open up a space in which to be more reflexive about the language of housing policy and the categories religiously employed in housing studies, such as the seemingly natural divisions between the 'market' and the 'state', or between the 'socially included' and the 'socially excluded' (Jacobs and Manzi, 2000).

In this chapter I want to explore how social constructionism can be used to critically interrogate contemporary policy metaphors that have captured the attention of housing researchers and policy makers in Australia and Europe. Specifically, I want to focus on the example of social exclusion and social inclusion as a way of illustrating the contested meaning of metaphors that inform housing policy and practices. Following Europe, social exclusion has emerged as an important theme in Australian housing policy debates (Arthurson and Jacobs, 2003). Rather than focus on how the concept can be measured and applied, I want to address the prior question about why this concept has emerged as a popular policy metaphor. What social relations and forms of knowledge have put social inclusion on the policy and political agenda? How does it achieve its effects on policy practices? I want to explore the argument that the homogenising tendencies embedded in some uses of the social exclusion/ inclusion discourse denies political contestation, obscures the relations that create social inequalities and clouds clear thinking about what could be done to create a more equitable housing system. The basis of my argument is that housing policy makers and housing researchers should be careful to avoid the narrow prescriptions that have plagued social inclusion discourse in related policy areas in Australia, such as income support and employment services.

A Cross-national Comparison of Social Exclusion Discourse

The ideas and practices informing social policy discourse in Western welfare regimes are regularly imported and exported between countries through a variety of formal and informal channels. The meaning of terms inevitably changes in this process of exchange, given the different institutional arrangements, political economy, history and culture that mediate the meaning and implementation of policy terms. The term 'social exclusion' is no different in this regard. As an umbrella concept, social exclusion is now used to represent a wide variety of social phenomena. It is variously employed by researchers and analysts in Europe and Australia to make connections between poverty, unemployment and underemployment, the spatial concentrations of socioeconomic disadvantage, housing estate regeneration and participation in political and cultural activities. The term is used to refer to a multitude of processes and features of the social order without being reducible to any one of them. There are, however, important cross-national differences in the social construction of the problem that social exclusion diagnoses and seeks to address. I do not intend to rehearse the full history of the contested emergence of social exclusion across national boundaries (see Byrne, 1999; Rodgers et al., 1995). However, it is useful to briefly sketch the contours of the meaning of the term, how it is different from poverty and inequality, and how meaning has changed over time and across space. This short history assists in addressing the question about the prominence of social exclusion and inclusion in Australian social policy circles.

The French Connection

The term social exclusion first emerged in France in 1974 and was used to refer to individuals not covered by the social security system. The use of the term in France originally had a structural meaning, referring to the impact of economic restructuring, in particular the decline in the manufacturing sector. The term was broadened in the late 1980s to encompass 'spatial concentrations of disadvantage', after a number of violent incidents on French social housing estates (Silver, cited by Arthurson and Jacobs, 2003: 3). In the 1980s, the term started to be used in France to refer to a process of social disintegration, in the sense of a progressive rupture of the relationship between individual and society because of increasing long-term unemployment, greater family instability and increasing numbers of homeless people (Rodgers et al., 1995).

The popularity of the social exclusion concept soon spread to neighbouring countries. Interest in social exclusion grew in Western Europe in relation to rising rates of unemployment, increasing international migration and cutting back of welfare state spending. The term was adopted by the European Commission in its mandate to report, on a European-wide basis, about prevailing levels of poverty and unemployment. As a description of a state of affairs, social exclusion closely relates to a state of poverty defined as 'relative deprivation', which refers to an idea that people cannot obtain the conditions of life (diet, amenities, standards and services) which allow them to participate as full members of society (Rodgers et al., 1995). In the political process associated with the formation of the European Union, social exclusion proffered a way of combating the problem whereby individual European Union member states could not agree on a proposed objective for combating poverty. As Marsh and Mullins (cited by Arthurson and Jacobs, 2003) state, social exclusion provided a way for 'member states to commit themselves to an imprecise, but nonetheless worthy-sounded mission'. Social exclusion is popular with politicians because the term is a shorthand way of calling for public action (Rodgers et al., 1995).

In Britain, the concept of social exclusion became formally wedded to a political programme with the election of New Labour in Britain in 1997. The problem of 'social exclusion' and the promise of 'social inclusion' became part of the political discourse in the lead up to the election, as a way of ensuring electoral success (Levitas, 1998). After winning the election, the UK Labour government established a 'whole of government' policy-making unit charged with the mission of tackling social exclusion. This defined social exclusion as:

> a shorthand term for what can happen when people or poor areas suffer from a combination of linked problems, such as unemployment, poor skills, low incomes, poor housing, high crime environments, bad health and family breakdowns (Social Exclusion and Cabinet Office, 2001: 2).

Much of the initial focus of the UK Labour government was on 'problem' public housing estates, as these were the areas that were seen to have a concentration of the problems identified above. This interest spurred on a number of reports and initiatives, including the report 'Bringing Britain Together: A National Strategy for Neighbourhood Renewal', which was released in 1998. The report was aimed at addressing the problems of the worst-off local communities. It asserted that poor housing problems are integrally linked to other problems, such as crime, unemployment and failing schools (Watt and Jacobs, 1999). In their analysis of the neighbourhood renewal strategy Watt and Jacobs (1999) argue that the meaning of social exclusion varied throughout the report; however, there was a tendency to define social exclusion in the report in terms of a moral underclass and social-integrationist discourse. The framework used by Watt and Jacobs was informed by the influential work of Ruth Levitas (1998) who identified three different meanings of social exclusion in British social policy. The three are:

> ... a redistributionist discourse (RED) developed in British critical social policy, whose prime concern is poverty; a moral underclass discourse (MUD), which centres on the moral and behavioural delinquency of the excluded themselves; and a social integrationist discourse (SID) whose central focus is on paid work (Levitas, 1998: 7).

In evaluating, the social exclusion discourse of the UK Labour government, Levitas (1998) suggests that the first term of the government was dominated by social integration discourse (SID) and a moral underclass discourse (MUD), rather than a redistribution discourse (RED). More recently, some British social policy scholars have suggested that the UK Labour government's social policies have started to give more attention to redistributionist policies through raising taxes for higher public spending, the provision of unprecedented tax credits for low-income people and establishment of the minimum wage (Bradshaw, 2003). While these achievements have been significant, the UK Labour government has also been quick to reassure voters that it is a government of low taxes and business values. The imperative to distance itself from previous Labour administrations continues to be politically important for the government. In Australia, low taxes and small government have been defining features of the Liberal/National Coalition government that came to power in 1996. In contrast to Britain, the take up of social exclusion discourse in Australia is somewhat patchy; it is less pronounced and less institutionalised in administrative programs. In Australia, the meaning of inclusion and exclusion is politically tied up with an abstracted notion of social cohesion, which is close to the social integration discourse (SID) identified by Levitas (1998).

A 'Socially Cohesive' Australia

Social exclusion permeates the spoken discourse of policy makers in Australia and it appears in a number of recent government consultation documents, however, these initiatives are not linked to any centrally driven or coordinating body. Nonetheless, social inclusion discourse in Australia has a certain rhetorical value in political discourse, linked to an idea of social cohesion and national identity. Australia's current Liberal-National government certainly considers social cohesion to be a crucial component of its policy platform. Prime Minister John Howard (18 September 2003: 1) has claimed that '[s]ocial stability and cohesion and caring for people is an integral part of Australian society. It's one of those things which is embraced by all sections of the community'. The term 'social cohesion' has not been explicitly defined by the Howard government, but political speeches and policy documents indicate that it is interpreted as being associated with 'belonging' and 'inclusion', which are represented as being based on shared values and norms (Humpage, 2003).

Social policy discourse at both the Commonwealth and state government level has become increasingly interested in enhancing social inclusion – often referred to in terms of social participation and levels of social capital – with community capacity building identified as a key policy tool with which to achieve these goals (Humpage, 2003). The Howard government's statement on welfare reform notes that:

> People who depend for long periods on income support rather than paid work face increased risk of financial hardship and social exclusion. The longer they spend out of work the harder it is to get another job and the more likely they are to lose confidence. This can have negative effects on their personal relationships and lead to a sense of detachment from society (Commonwealth of Australia, 2002: 5).

Here we see how social exclusion is associated with 'welfare dependency', with the principle pathway out of social exclusion defined in relation to 'paid work'. In 'workfare' orientated countries (USA, more recently the UK and now Australia) paid work is taken to be the principal gauge of 'social inclusion' and is seen as promoting 'self-reliance' (Family and Community Services, 2003). This particular policy narrative avoids concerns raised by welfare organisations and academics about the problems with precarious employment and the low-wage end of the labour market. Through its policies and programs, the Commonwealth government remains intent on getting people into any form of employment, which in practice risks simply replacing one form of poverty with another – the only difference being that one is called 'self-reliance' and the other is pejoratively cast as 'welfare dependency'. The single-minded focus on paid work as the only recognised form of participation fails to recognise the importance of non-economic activities in the formation of social citizenship.

The focus on paid work ignores and devalues a range of other activities such as unpaid work, volunteering, caring and parenting.

Defined in relation to inclusion in the labour market and job participation, concepts such as 'social inclusion' are now used to compel citizens and justify mutual obligation requirements for beneficiaries which have been portrayed as part of the social contract that citizens have with the liberal democratic state (see Moss, 2001: 2; Harris and Williams, 2003: 206). The over-arching vision of social policy held by the current Commonwealth government emphasises notions of social coalition and community-business partnership to facilitate stronger families and communities (Stone, 2001: 6). Community capacity building initiatives are regarded as crucial to meeting such goals because they have aimed to 'encourage the development of community capacity for self-help; help ameliorate the effects of pressures on and within communities; facilitate partnerships between business, community groups and governments to achieve well-targeted and tailored solutions' (Family and Community Services, 2003).

Through these initiatives individuals are encouraged to manage their own welfare in the interests of establishing ethical communities. The deployment of community is used to identify a territory between the authority of the state, the free and amoral exchange of the market and the liberty of the rights-bearing individual (Rose, 1999: 5). Within this space, citizens are constructed as free from government instrumentalities that are increasingly associated with an oppressive and disabling state. Thus, community becomes a counterweight to politics (Marston and McDonald, 2002: 386). This romanticisation of community fails to recognise that there will always be characteristics of communities that include people and exclude others. In sum, the advocacy of community may be regarded as an attempt to grapple with some of the problems of social exclusion, but it provides no guarantee that sources of exclusion will be eradicated (Little, 2002: 146).

In addition to a renewed interest by policy makers in ideas such as 'community' and 'inclusion', sections of the academic social policy community in Australia have also appropriated the language of social exclusion and social inclusion.[1] The Australian Social Policy Conference 2003, for example, was simply titled *Social Inclusion*. Somewhat surprisingly for an academic conference there was very little space devoted in the conference program to critically interrogating the utility and meaning of social inclusion and social exclusion. Peter Saunders (2003) has explored this question in some detail through a review of social exclusion in Australia to determine what it adds to poverty research. Saunders (2003: 17) concludes by arguing that:

> Overall, there are two major barriers preventing social exclusion exerting anything more than a marginal influence on policy development in the foreseeable future. The first is a lack of will, or interest, among key agencies or individuals in the Howard Government. The second is the lack of any clear common interest in tackling the causes and consequences of social exclusion between the Commonwealth and State Governments.

Saunders suggests that it is a question of political will and a consequence of Australia's federated system that has led to a conservative approach to processes of inclusion and exclusion. Saunders' main focus and interest in social exclusion is with the appropriateness of various social and economic indicators that could be used to measure whether individuals and social groups are 'socially excluded'. Unfortunately, Saunders spends less time addressing the basic issue about whether the concept has any empirical referent. The problem with a preoccupation with methodology and measurement is that the concept of social exclusion is more or less appropriated by 'number crunchers' (statisticians, economists and empirical sociologists) before its utility and value is thoroughly explored. This issue is not unique to the concept of social exclusion, but is endemic to a model of the social sciences that attempts to ape the physical sciences. This serves to underline the point that how we think and talk about social exclusion/inclusion cannot be divorced from a discussion about the role played by the social sciences in the constitution of the 'social'. In the next section I want to briefly discuss the politics of policy making in the framing of policy problems, such as social exclusion.

Rethinking the Task of the Social Sciences

In the development of techniques of social discovery and social policy-making, social scientists have traditionally seen themselves as experts and as spectators deploying the methods of the natural sciences to discern the underlying laws and structure of 'the social'. As Manent (1998: 50–85) has argued, the evolving social sciences of the late nineteenth century (like economics, psychology and sociology) formulated their idea of a 'science of man' with paradoxical effects, chief of which was to obscure any real interest in the nature of human conduct. Manent (1998) argues that whereas classical political and philosophy had adopted the viewpoint of the practical actor as citizen; social scientists early took the 'viewpoint of the spectator'. The viewpoint of the spectator is 'pure and scientific' in that it accords no real initiative whatever to the human actor/s but considers their action or works 'as the necessary effects of necessary causes. In this paradigm, policy research is defined as an exercise in instrumentally identifying and measuring taken for granted problems (such as unemployment) and then identifying a range of possible solutions (active labour market programmes, for example, or the reduction of poverty traps).

Flyvbjerg (2001) suggests that it is high time we redirected the orientation of the social sciences away from a universal and rational grounding and move towards a socio-historical foundation concerned with power relations, the production of exemplary case studies that get close to reality and which reconnect agency and structure. Along the same lines, Flyvbjerg (2001) encourages social scientists to pose questions about power, outcomes and inequality in terms of: who gains and who loses? What possibilities are available to change existing power relations? What kinds of power relations are those asking these questions themselves a part of? The emphasis here is on dynamic *how* questions, rather than structural *why* questions.

The priority given to *how* questions includes a focus on how social groups are constructed by discourses and constituted in concrete practices. We need to recognise that the very idea of inequality and exclusion is implicated in the politics of naming people and problems. The way people or problems are named is absolutely crucial, because policy making is an irreducibly linguistic and political process. 'Problems' or 'issues' only come to be that way when they have become part of a policy, or political discourse. Television news and current affairs, talkback radio, community submissions and opinion polls are all popular discursive forms where housing policy messages are communicated by those who claim the authority to name and represent. At different points in time, we have politicians, social commentators and the media warning us about 'spiralling housing costs', 'interest rate rises', 'social exclusion', 'overcrowding' and 'tenants from hell'.

In whipping up a moral crisis, the discursive construction of social issues does the job of selling as well as telling. The words used to represent and diagnose the problem are critically important in policy; however, discourse is a field of social and political action that often escapes the close attention of policy researchers. The symbolic side of policy deserves more attention than it often receives in conventional social inquiry. As Majone (cited by Fischer and Forrester, 1993: 2) argues: 'As politicians know only too well, but social scientists often forget, public policy is made of language'. Not only do discursive meanings make the social welfare state, they are a necessary part of the explanation of the social and political actions of the citizens within it (Fischer, 2003).

Governments and policy communities (politicians, party officials, civil servants and experts and academics across various scientific and technical backgrounds) both constitute problems, and solutions to those problems. Yeatman (1990: 160) reminds us that policy making is pre-eminently a politics of contest over meaning. This is because it: 'comprises the disputes, struggles and debates about how the identity of the participants should be named and thereby constituted, [and] how their relationships should be named and thereby constituted' (Yeatman, 1990: 160). The naming of a person or group in a particular way situates them in regard to social, legal or economic practices and entitlements. Van Dijk (1998), for example, argues that the term 'illegal refugees' is employed in political discourse to induce hostility towards a group that has supposedly 'broken the law', which mitigates against a compassionate concern for a person's welfare. The construction of social identities shapes people's capacities for mobility, for example, as the occupants of the 'business class', as 'genuine refugees', as 'the homeless' or as 'unwanted dependents' and as 'economic migrants' (Clarke, 2000: 209).

Peel (2003) reminds us that those people habitually referred to as 'the poor' or 'the socially excluded' do not name themselves in this way. Peel's (2003: xi) research project involving interviews with hundreds of public housing tenants on broad acre public housing estates is very revealing in this respect:

> If those to whom I spoke were best characterized as disadvantaged, they mostly called themselves 'ordinary'. Some people preferred the word 'battlers', though they were growing suspicious of a term twisted by conservatives to mean people who had more than they had. A few – normally those with strong union or Labour party backgrounds – used terms such as 'working class'. In general though, explicit class language was reserved for others.

Though Peel (2003: xii) claims that, 'There is as yet no agreed terms for Australia's poorest citizens', there is nonetheless a rough kind of consensus of meaning at work when so many Australians speak about 'the poor', 'the homeless' and increasingly about 'welfare dependents'. The persuasive power of policy metaphors, such as social exclusion and social inclusion are able to operate as effective descriptions of social reality because they address an emotional and intuitive appeal to 'common sense'.

Harris and Williams (2003: 206) refer to this neglected dimension of policy analysis as the moral imagination – that is social inclusion deals with how participatory relations *ought* to operate, relies heavily on *representation*, and has *emotional* (rather than rational-calculative) foundations. The popular conception of social inclusion in Australian social policy discourse, for example, tends to smooth over competing conceptions about what constitutes the 'social' because it holds values and norms as consensus-based, fixed and universal. The way social inclusion is left unspecified in many policy documents is a reflection of the fact that we supposedly all subscribe to a popularised 'Australian way of life' that encompasses home-ownership aspirations, paid work and hetero-normative notions of care (see, for example, Commonwealth of Australia, 2002). These assumptions strike at the very heart of what is problematic about social exclusion.

What is 'Social' about Exclusion?

To properly evaluate a term like 'social exclusion' we need to grapple with the contested meaning of the 'social' in late modernity. The starting point is to ask: what do we mean when we talk about the 'social'? What are people excluded from? What sort of collective community is being imagined in these descriptions? Is it multiple or singular, consensus driven or conflictual? To what extent is the social something that is written and spoken? Answering these questions requires a recognition that the 'social' is something that is accomplished, produced and defined in everyday interactions between citizens and forms of government; it is socially constructed in the contested borders between private concerns and public matters. Recognising that the social is an historical artefact unsettles the idea that the 'social' refers to something objective and static.

A thorough exploration of what constitutes the social is beyond the scope of this chapter. My intention is to simply put forward the provocation that the question about what constitutes the 'social' is always worth asking. History, context, practice and texts are all important variables that constitute what the 'social' means. What counts as 'social' as a focus of social policy matters deeply for members of a given society (Lewis, 2000). Whether unemployment is a focus of social policy; whether childcare is considered a private or public responsibility, whether refugees are embraced by nation states or expelled at the borders; and whether housing is understood as a basic human right or a private responsibility has significant consequences for what we mean by social citizenship. One dimension of adopting a constructionist approach is the challenge to conventional assumptions about what is – and what is not – social.

A short history of the 'social' reminds us that there is nothing pre-given about what we commonly refer to as society. Pat O'Malley (1996: 2) has well summarised the way the idea of 'the social' came into play through the eighteenth and nineteenth centuries. 'Society' and 'the social' were constituted:

> ... in terms of a collective entity with emergent properties that could not be reduced to the individual constituents, that could not be tackled adequately at the level of individuals and that for these reasons required the intervention of the state. Social services, social insurance, social security, the social wage were constituted to deal with social problems, social forces, social injustices and social pathologies through various forms of social intervention, social work, social medicine and ... [social policy].

These various categories led to the eventual establishment of a virtual army of experts on the 'social', including sociologists, social planners and social workers. These were able to deploy a range of techniques to measure and manage the collective whole and the individual. It was the invention of statistics and the birth of 'the population' that led to increasingly detailed knowledge about family characteristics, individual productivity, wealth, death rates and illness. In the eighteenth century the 'family' was subsumed within the population and the economy and relegated to a secondary importance, except of course when it came to collecting statistics and as a site of moral and religious instruction. The social sciences are riddled with the formation of such objects of knowledge: our 'humanity', the 'economy', the 'welfare state' and the 'child' are all examples familiar to the social sciences.

In sum, the category of 'society' emerged and became an organic entity that could be measured in terms of its overall health, deterioration, competitiveness and even integration with other national societies. From this perspective, 'the social' does not refer to an inescapable fact about human beings – that they are social creatures – but to a way in which human intellectual, political and moral authorities, within a limited geographical territory thought about and acted upon their collective experience (Rose, 1999: 101).[2] It is the science of the social that gives rise to the very possibility of measuring degrees of exclusion and inclusion within this geographical and moral territory. One of the risks in adopting an adopting a moral binary approach to social inclusion, however, is that policy makers may come to accept the rational belief in the possibility of a consensus based model of a public sphere.

The social exclusion literature is beginning to take account of the spatial and relational dimensions of social inequality (Fincher and Saunders, 2001). To date, however, this analysis has largely failed to encompass the public spaces and spheres of communicative action where the social identities of 'the poor', 'the homeless', 'tenants from hell' and 'welfare dependents' are constructed and contested. This communicative space deserves more recognition, given that this is where legitimate action and inaction on the part of governments, community agencies, citizens and social scientists is constructed and contested.

My intention is not to rehearse the history of the role of the public sphere in the modern liberal tradition. I simply want to flag the relationship between material disadvantage and participation in public and cultural life as an increasingly important, but often unrecognised dimension of social exclusion discourse. Recent interest in the democratic public sphere was sparked when the English version of Jürgen Habermas's book the *Structural Transformation of the Public Sphere* was published in 1989 (Fraser, 1995). In this book and in his later work, Habermas has promoted the transforming and democratic potential embedded in a deliberative mode of communication between citizens. For Habermas, the public sphere is a site where differences are respected and where inequalities are bracketed as citizens participate as social equals in the practices of reasoned argument. Habermas's conception of a single public sphere, however, ends up sounding idealistic and utopian. As Fraser (1995: 289) argues:

> Informal impediments to participatory parity can persist even after everyone is formally or legally licensed to participate. When this happens deliberation can mask domination ... Even the language people use as they reason together usually favours one way of seeing things and discouraging others. Subordinate groups sometimes cannot find the right voice or words to express their thoughts, and when they do, they discover they are not heard. Thus, social inequalities tend to infect deliberation, even in the absence of formal exclusions.

The above critique raises serious problems with the modern liberal conception of a public sphere and its potential to address social exclusion. Moreover, the emergence of new social movements challenges the possibility and utility of a single public sphere. In fact, it is the persistence of social inequalities that gives rise to counter public spheres as a space where claims are made by communities of interest for recognition and the redistribution of resources and services. With a conception of multiple publics we can comprehend the relations of dominance and resistance, participation and antagonism, and thereby explain the co-appearance of the labour movement in relation to the development of capitalism, free speech movements in relation to the control of information and feminism and gay and lesbian rights movements in relation to patriarchal heterosexism (Gagnier, 1991: 10). These developments recognise the multiple lines of struggle and difference that characterise contemporary societies.

In sum, we need an analytics of power that renders visible the ways in which social inequality taints deliberation and participation within existing public spheres and how inequality affects relations among different social groups in such societies (Fraser, 1995). This framework points to a need for a more expansive and encompassing approach that recognises both the cultural and material dimensions of injustice. Nancy Fraser (1997: 15) eloquently draws attention to the relationship between cultural and economic justice:

> Cultural norms that are unfairly biased against some are institutionalised in the state and the economy; meanwhile, economic disadvantage impedes equal participation in the making of culture, in public spheres and in everyday life. The result is often a vicious circle of cultural and economic subordination.

This analysis points to a conception of social wellbeing that includes opportunities for participation in public life, access to policy discourses and other cultural activities. A construction of social exclusion that includes political rights and citizenship helps us go beyond economic and social aspects of poverty and inequality. In this sense, the term 'social exclusion' can potentially be a value-adding concept, depending – of course – on how it is constructed.

Implications for Policy Practice

As noted earlier, social inclusion discourse in Australia is preoccupied by the promotion of paid work, which idealistically becomes the basis of inclusion in a singular moral community that upholds the ethic of work and self-reliance. Social integration is represented as a process of bringing 'people on the margins' into a seemingly homogenous centre. By definition, social exclusion carries with it the imperative of inclusion, it is not happy with the excluded being outside the ranks of social citizenship (Young, 2003). This tension between inclusion and exclusion brings with it a certain 'will to govern' on the part of policy makers, which in practice can manifest as the assertion of a one-dimensional voice in attempting to deal with social concerns.

We have also noted that the 'social' is increasingly constructed as a site of 'community'. In social exclusion discourse this 'community' can have a geographical boundary (as in the case of a degenerated public housing estate) or it can be defined in terms of social identity, as a community of interest (the worker, the tenant, the underclass). In contrast to social exclusion, which can be defined by a multitude of differences, the imperative in social inclusion rhetoric is to get marginal communities to overcome their differences and 'sign up' to 'society'. Little (2002: 175) argues that this approach to inclusion and exclusion is doomed to fail:

> The greatest weakness of political communitarianism is its failure to recognise that the community that it constructs as the key agent of the cohesive society is a fallacy. Imposing homogeneity or universal theories of rationality detaches the political commentator from the realities of plural society.

In contrast to this homogenising model, policy makers need to recognise differences and the likelihood of conflict and dissent (Little, 2002: 147). In practical terms, for example, this means that public housing managers need to acknowledge the inevitably of neighbourhood disputes between tenants and home-owners in urban spaces defined by contradictory needs and desires. And in terms of crime, it means recognising that a repeat offender may feel more 'included' in prison among their peers than they do participating in mainstream social activities such as paid work. These realities cannot be dismissed. Social antagonism and dissensus cannot simply be wished away in the naïve and fallacious search for the grand narrative of universal social inclusion. Processes of inclusion and exclusion are imbued with relations of power and these relations can only be properly understood in specific contexts and practices.

In short, there is no social, economic, cultural or political order to which we all subscribe. Accordingly, we need a radical conception of community that recognises conflict, or what Mouffe (2000) refers to as 'agonistic pluralism' – defined as an acceptance that power is unavoidable. Mouffe's thesis is that it is not disagreement *per se* that inhibits contemporary democracies, but rather, the antagonistic fashion in which such differences manifest themselves. Mouffe's approach advocates an idea of conflict as a relation between adversaries (rather than enemies) and as such seeks to addresses the problem embedded in orthodox communitarianism, which pretends that confrontation can be overridden by the dominant morality (Little, 2002). This critique of universalism should not be equated with relativism. What I am asserting is the importance of 'contextualism' and proper attention to concrete practices and discursive politics (Flyvbjerg, 2001).

It is possible to advocate social transformation without putting forward a universal blueprint for the future, or to impose a view of the world that neatly divides people into the socially excluded and the socially included. We need to remind ourselves that social exclusion and social inclusion are categories invented by social scientists and politicians. This is not to deny that there are cases and experiences of objective poverty. Regrettably there are too many cases, even in the twenty-first century, where people die and/or are dying slowly because they lack the barest necessities of life, like a supply of food, shelter and water such as would keep them alive. None of what I have written should be read as ignoring the lived experience of those who experience inadequate income, pain, misery, homelessness and eviction, distress, cold, and hunger. Nor do I deny the lived reality of reactions like frustration, anger or humiliation in response to well-meaning community workers or welfare bureaucrats that so many low-income people can report. The problem is the continual attempts to invest the idea of 'poverty', 'social exclusion' and 'the poor' with universal meanings. As Dean (1999) powerfully reminded us, 'poverty' is an idea that has come out of a long discursive history.

What I am advocating is a reflexive approach to the social categories used by social scientists and policy makers to diagnose and address social concerns. An ongoing and important task of the social scientist and the policy maker alike is to interrogate the preconstructed categories of 'immigrant populations', 'youth at risk' or 'long-term unemployed' and 'the homeless'. Our knowledge about the social world invariably informs the sorts of policies that we try and inscribe upon it in order to shape it and twist it, hence the importance of reflexivity. As Bourdieu (1992: 107) explains:

> Reflexivity is a tool to produce more science, not less. It is not designed to discourage scientific ambition but to help make it more realistic. By helping the progress of science and thus the growth of knowledge about the social world, reflexivity makes possible a more responsible politics.

I am not suggesting that policy makers and housing activists do away with claims for better access to social, economic and cultural resources, but we need a language of specificity, not umbrella concepts that lack a concrete form of analytics or modality. I agree with Arthurson and Jacobs (2003) in their review of social exclusion and housing research that social exclusion is inadequate as an academic concept and offers no advantage over other commonly used ideas such as 'housing related poverty'. Arthurson and Jacobs (2003) suggest that social exclusion may be valuable as an idea that can be used to link up related processes of economic and social disadvantage, such as the need to coordinate housing policy with investment in education and transport, employment and training and crime prevention. Implicit in this definition of social exclusion, however, is an emphasis on the need for investment and distribution of resources to address social concerns. This more *structural* diagnosis of social exclusion is very different to an *agency* orientated discourse that associates exclusion with social integration and social policies that focus on managing the conditional citizenship rights of the 'moral underclass'. All of which simply underlines the point that social exclusion is an extremely slippery concept.

In discussing and debating the construction of social concerns we may need to look to alternative policy metaphors that highlight rather than obscure material deprivation and patterns of inequality. While social exclusion and 'community capacity building' acknowledge the importance of place and space, they run the risk of 'othering' ordinary people. Simply applying the label of social exclusion to geographical areas of concentrated disadvantage is insufficient. As Arthurson and Jacobs (2003: 26) argue:

> The literature demonstrates that using social exclusion in this way is labelling the symptoms, rather than using social exclusion as a tool to understand the processes of decline, which adds nothing to policy debates about housing and inequality and can add to the stigma of these localities.

In continuing to use these popular metaphors, the tone of government social policy discourse runs the risk of being influenced by the fusion of conservative communitarianism, underclass sentiments and the relegation of social justice to a largely rhetorical sphere (Harris, 2000). Moreover, policy metaphors and meanings need to be culturally appropriate. European literature on social exclusion, for example, tends to associate poverty and deprivation as associated with 'lack of integration'. In contrast, in Latin America poverty is seen as being structurally related to the ways economies and societies function, which are grounded in the peripheral integration of Latin American economies into the world capitalist system (Rodgers et al., 1995: 5).

In light of this critique, there is nothing universally sound or scientific about the concept of social exclusion. As such, we need to be cautious about using 'big ideas' if they are appropriated and used to disinvest 'communities' of resources in the name of promoting self-reliance, social order and social participation. We must be careful to avoid a situation where the problem of what to do about poverty becomes a problem of what to do about the poor (Peel, 2003). The gaze of the social sciences and policy makers should be careful to avoid a preoccupation with governing the values, attitudes, practices of 'the poor' in the name of building social inclusion. Inequalities in the housing system, for example, are not primarily the result of tenants lacking motivation, capacities, resilience, skills and determination. Our present housing system is a consequence of historical choices that have been made by successive governments in Australia about where to channel resources. Housing related poverty and homelessness in Australia should therefore not be conducted as a debate about scarcity – it is about political choices. The predominance of home-ownership in Australia, for example, is not a reflection of some inherent cultural value; it is a reflection of a housing system that has historically favoured home-ownership through a range of taxation measures, incentive schemes and weak provisions for security of tenure and affordability in the private rental market.

There are no shortage of practical proposals and principles to address these inequities. Yet, contested choices and political decisions can be obscured by social inclusion talk that narrowly and conservatively focuses on the ethic of paid work, the problem of welfare dependency and a representation of 'the poor' as lacking the required human capital to get ahead. This approach ignores the barriers and forces that reproduce relations of economic, cultural and social inequality. If ordinary people are to be held to account for their social obligations to each other in the name of the collective good and nation building than so must those rendered extraordinary by their wealth, their power, their luck or their talent (Peel, 2003: 175).

Attention to multiple perspectives on the relations of social inequality suggests that we should pay close attention to the social categories invoked by those that are the silent targets of social exclusion policies. Social researchers and policy-makers need to be ever mindful of what it means to practice a politics of listening and respect. It is a form of 'cultural injustice' (Fraser, 1997), for example, to suggest that public housing tenants living on degenerated public housing estates are somehow different from the rest of us – a group apart – lacking certain moral virtues. This form of social positioning in the processes of demarcating between the 'included' and the 'excluded' is referred to by Bourdieu (1992) as a form of 'symbolic violence'. Symbolic violence can be seen in identity constructions that fix upon a partial social categorisation and impute it to a limiting set of moral characteristics (Taylor, 1998: 342). *Access* to the discourse practices of policy making is an important intervention in challenging these prescriptions and reclaiming respectful identities. Without this possibility those who do not have the 'means of speech', or do not know how to 'take the floor' can only see themselves in the words or discourse of others – that is those who are legitimate authorities and who can name and represent (Mahar et al., 1990: 15).

Conclusion

So where does all this leave us? The discussion has highlighted the variety of ways that social exclusion can be constructed in policy discourse, which in large part depends on the model of the social that is imagined and acted upon. There is a very big difference between understanding social relations in a rational, consensus-based model, or more radical notions that accept the inevitably of social conflict. These foundational questions deserve more attention than they currently receive in policy discussion about the utility of social exclusion and social inclusion. These policy metaphors need not be accepted at face value. We need to recognise the value in a multiplicity of conceptual and methodological approaches. Poverty, for example, is an ideological formation, it is a truth produced by particular discursive strategies, it is also a social construction – and people die from it (Clarke, 1998: 183).

I readily acknowledge that there is always a risk in wholeheartedly abandoning conventional paradigms and embracing the new. In using a social constructionist approach in this chapter I am not suggesting that all social practices be reduced to discourse, or that social constructionism replace other epistemologies. There is nothing particularly novel or enlightening about the argument that social reality is constructed. The value of the analytical method associated with social constructionism comes from showing how concepts and practices are constructed, to make policy 'truths' appear less self-evident and to open up the possibility that such practices could be constructed differently with different effects. To open up these possibilities will require more encompassing language about the nature, extent and consequences of social inequality and injustice, an analytics of inequality that encompasses measurement and meaning, rationality and values, processes and structures.

The homogenising tendencies built into mainstream definitions of social inclusion and exclusion (and similar social policy metaphors, such as social capital) means that we may have to look elsewhere for concepts that capture the multiple lines of difference that characterise contemporary societies. It makes no practical sense to talk in general terms about social inequality or social exclusion, given that these terms lack conceptual specificity. People in social contexts do not exist one-dimensionally as 'poor people', 'black people', as 'women' or as 'people with disabilities'. The unequal distribution of wealth, income and cultural resources flows through a variety of social categories that make a difference to the place of individuals (Clarke, 2000: 208). In recognising these multiple lines of distinction we need to move beyond universal conceptions of social inclusion and exclusion.

Addressing these multidimensional concerns requires an expanded 'social policy imaginary' (Lewis, 2000) and a regenerated public sphere – as a means of reclaiming disrespected identities and revaluing a more encompassing definition of social citizenship. Treating people with respect, however, cannot simply be commanded. As Sennett (2003: 260) concludes: 'Mutual recognition has to be negotiated; this recognition engages the complexities of personal character as much as social structure'. Such negotiations need to occur at a macro and micro level, in everyday interactions and at the level of political discourse. The heart of Sennett's (2003) thesis is that the key political issue for contemporary welfare states is not the continuing existence of social inequality, poverty and exclusion *per se*, but how difference and autonomy are treated in everyday social relations between government agencies and citizens, between doctors and patients, between Centrelink officers and clients, between public housing managers and tenants.

Notes

1   Arthurson and Jacobs (2003) note that despite misgivings about social exclusion as an explanatory concept, academics in the UK had little choice about engaging in these debates because the UK Labour government put all action against disadvantage under the banner of social exclusion.
2   What is interesting about the social policy discourse in the contemporary era is that the 'social' is being replaced by a range of other equally amorphous terms. In 1996, for example, the Australian government abolished the *Department of Social Security* and established *Centrelink* to administer income support. The idea of being 'on social security benefits' has been thoroughly problematised and the stated aim of governments in countries such as Britain, Australia and America is precisely to get citizens 'off the social' (Rose, 1999: 100). This is best expressed by the notion of 'welfare-to-work'.

References

Arthurson, K. and Jacobs, K. (2003), *Social Exclusion and Housing*, Melbourne: Australian Housing and Urban Research Institute.
Bacchi, C. (1999), *Women, Policy and Politics*, London: Sage Publications.
Bourdieu, P. (1992), 'The Purpose of Reflexive Sociology', in P. Bourdieu and J. Wacquant (eds), *An Invitation to Reflexive Sociology*, Chicago: University of Chicago Press: 62–213.
Bradshaw, J. (2003), *How Has The Notion of Social Exclusion Developed In The European Discourse?*, Plenary Address, Sydney: Australian Social Policy Conference.
Byrne, D. (1999), *Social Exclusion*, Buckingham: Open University Press.
Clarke, J. (1998), 'Thriving on Chaos? Managerialism and Social Welfare', in J. Carter (ed.), *Postmodernity and the Fragmentation of Welfare*, London: Routledge: 101–15.

Clarke, J. (2000), 'A World of Difference: Globalisation and the Study of Social Policy', in G. Lewis, S. Gewirtz and J. Clarke (eds), *Rethinking Social Policy*, London: Sage Publications.

Commonwealth of Australia (2002), *Building a Simpler System to Helpless Jobless Families and Individuals*, Canberra: Commonwealth of Australia.

Dean, M. (1999), *Governmentality: Power and Rule in Modern Society*, London: Sage Publications.

Family and Community Services (2003), '*What Makes Strong Families and Strong Communities*', <http://www.facs.gov.au/internet/facsinternet.nsf/whatfacsdoes/families-FamilyAssistance.htm>.

Fincher, R. and Saunders, P. (2001), 'The Complex Contents of Inequality', in R. Fincher and P. Saunders (eds), *Creating Unequal Futures: Rethinking Poverty, Inequality and Disadvantage*, Sydney: Allen and Unwin: 1–15.

Fischer, F. (2003), *Reframing Public Policy: Discursive Politics and Deliberative Practices*, Oxford: Oxford University Press.

Fischer, F. and Forrester, J. (eds) (1993), *The Argumentative Turn in Policy Analysis and Planning*, Durham: Duke University Press.

Flyvbjerg, B. (2001), *Making Social Science Matter: Why Social Inquiry Fails and How it Can Succeed Again*, Cambridge: Cambridge University Press.

Fraser, N. (1995), 'Politics, Culture and the Public Sphere: Towards a Postmodern Conception', in L. Nicholson and S. Seidman (eds), *Social Postmodernism: Beyond Identity Politics*, Cambridge: Cambridge University Press: 287–314.

Fraser, N. (1997), *Justice Interruptus: Critical Reflections on the 'Postsocialist' Condition*, London: Routledge.

Gagnier, R. (1991), *Subjectivities: A History of Self-Representation in Britain, 1832–1920*, New York: Oxford University Press.

Hacking, I. (1999), *The Social Construction of What?*, Cambridge: Harvard University Press.

Harris, P. (2000), 'Participation and the New Welfare', *Australian Journal of Social Issues*, Vol. 34, No. 5: 279–92.

Harris, P. and Williams, V. (2003), 'Social Inclusion, National Identity and the Moral Imagination', *The Drawing Board: An Australian Review of Public Affairs*, Vol. 3, No. 3: 205–22.

Hastings, A. (1998), 'Connecting Linguistic Structures and Social Practices: A Discursive Approach to Social Policy Analysis', *Journal of Social Policy*, Vol. 27, No. 2: 191–211.

Howard, J. (2003), Address to the Queensland Liberal Party State Convention, Gold Coast, 14 September, <http://www.pm.gov.au/news/speeches/index.cfm?speechYear=2003>.

Humpage, L. (2003), 'Throwing the Baby Out with the Bathwater: Contradictions in Government Policy Regarding Refugees on Temporary Protection Visas, Social Inclusion and Social Cohesion in Australia', paper presented to the *Sociological Association of New Zealand Annual Conference*, Auckland, 9–11 December.

Jacobs, K. and Manzi, T. (1996), 'Discourse and Policy Change: The Significance of Language for Housing Research', *Housing Studies*, Vol. 11, No. 4: 543–60.

Jacobs, K. and Manzi, T. (2000), 'Evaluating the Social Constructionist Paradigm in Housing Research', *Housing, Theory and Society*, Vol. 17, No. 1: 35–42.

Levitas, R. (1998), *The Inclusive Society: Social Exclusion and New Labour*, Basingstoke: Macmillan.

Lewis, G. (2000), 'Expanding the Social Policy Imaginary', in G. Lewis, S. Gewirtz, and J. Clarke (eds), *Rethinking Social Policy*, London: Sage Publications: 1–22.

Little, A. (2002), *The Politics of Community: Theory and Practice*, Edinburgh: Edinburgh University Press.

Mahar, C., Harker, R. and Wilkes, C. (1990), 'The Basic Theoretical Position', in C. Mahar, R. Harker and C. Wilkes (eds), *An Introduction to the Work of Pierre Bourdieu: The Practice of Theory*, London: The Macmillan Press: 1–16.

Manent, P, (1998), *The City of Man*, trans. M. Le Pain, Princeton: Princeton University Press.

Marston, G. (2000), 'Metaphor, Morality and Myth: A Critical Discourse Analysis of Public Housing Policy in Queensland', *Critical Social Policy*, Vol. 20, No. 3: 349–73.

Marston, G. and McDonald, C. (2002), 'Patterns of Governance: The Curious Case of Non-profit Community Services in Australia', *Social Policy and Administration*, Vol. 36, No. 1: 376–91.

Marston, G. and Watts, R. (2003), 'Tampering with the Evidence: A Critical Appraisal of Evidence Based Policy', *The Drawing Board: An Australian Review of Public Affairs*, Vol. 3, No. 3: 142–63.

Moss, J. (2001), 'The Ethics and Politics of Mutual Obligation', *The Australian Journal of Social Issues*, Vol. 36, No. 3: 1–14.

Mouffe, C. (2000), 'For an Agonistic Model of Democracy', in N. O'Sullivan (ed.), *Political Theory in Transition*, London: Routledge.

O'Malley, P. (1996), 'Criminology and the New Liberalism', *The 1996 John Edwards Memorial Lecture*, <http://criminology.utoronto.ca/edwardslect/>.

Peel, M. (2003), *The Lowest Rung: Voices of Australian Poverty*, Melbourne: Cambridge University Press.

Rodgers, G., Gore, C. and Figueiredo, J. (1995), *Social Exclusion: Rhetoric, Reality, Responses*, Geneva: International Institute for Labour Studies.

Rose, N. (1999), *Governing the Soul: The Shaping of the Private Self*, London: Free Association Books.

Saunders, P. (2003), *Can Social Exclusion Provide a New Framework for Measuring Poverty*, Sydney: Social Policy Research Centre.

Sennett, R. (2003), *Respect: The Formation of Character in an Age of Inequality*, London: Allen Lane.

Social Exclusion and Cabinet Office (2001), 'What is Social Exclusion', <http://www.cabinet-office.gov.uk/seu/index/march_%202000_%20leaflet.htm>.

Stone, W. (2001), *Measuring Social Capital: Towards a Theoretically Informed Measurement Framework for Researching Social Capital in Family and Community Life*, Research Paper No. 24, Melbourne: Australian Institute of Family Studies.

Taylor, D. (1998), 'Social Identity and Social Policy: Engagements with Postmodern Theory', *Journal of Social Policy*, Vol. 27, No. 3: 329–50.

Van Dijk, T. (1998), *Ideology*, London: Sage Publications.

Watt, P. and Jacobs, K. (1999), 'Discourses of Social Exclusion: An Analysis of 'Bringing Britain Together: A National Strategy for Neighbourhood Renewal', paper presented at the *Discourse and Policy Change* Conference, University of Glasgow 3–4 February.

Yeatman, A. (1990), *Bureaucrats, Femocrats, Democrats: Essays on the Contemporary Australian State*, Sydney: Allen and Unwin.

Young. J. (2003), *Social Exclusion*, <http://www.malcolmread.co.uk/JockYoung/social_exclusion.pdf>.

Chapter 6

Housing Pathways – A Social Constructionist Research Framework

David Clapham

The aim of this chapter is to illustrate the usefulness to housing research of the concept of a housing pathway. The pathways approach involves the application of a metaphor of a pathway to the movement of a household between different housing situations through the lifecourse. Although the pathways metaphor could in principle be used independently of social constructionism, in its recently developed form it is deeply embedded in the social constructionist paradigm (Clapham, 2002). It fulfils the need for a bridge between the overall social constructionist paradigm and its application to the specific field of housing.

Social constructionism is a broad paradigm with many differences of emphasis within its parameters. Nevertheless there is a core of basic tenets which are briefly described in the first section of the chapter. However, it is argued that social constructionism needs to 'borrow' concepts from outside the tradition if it is to have the conceptual armoury for an analysis of housing. Particularly useful is the idea of 'structuration', which links human agency to structural opportunities and constraints. Also important is a conceptualisation of power.

The aim of the next section is to give a flavour of the pathways approach by exploring four key dimensions. This is done by using the approach as an analytical framework to describe existing knowledge in the area and to identify gaps which a pathways framework would expose. The discussion of each dimension contains an outline of an important discourse which frames action and includes a discussion of existing knowledge of household meaning and behaviour. This structure reflects the contingent action emphasis in the pathways approach.

The first key issue is the form and workings of the household, which is the cornerstone of the approach because it is the unit in which people consume housing and make decisions about that consumption. The composition and workings of the household are influenced by discourses of the family which structure political debate and frame the decisions of households. The second key area of analysis is employment or work. The discourse of a flexible labour market has strongly influenced housing policy. At the same time the nature of employment has influenced the housing decisions of individual households. The third key area is the way that households pay for housing. Discourses have shaped government intervention in the institutional structure of housing finance which households have to navigate in order to pay for their housing. The example given here is the discourse of globalisation. Within this institutional structure households make decisions about how much to spend on housing, although we know little about the decision-making processes involved. The fourth area is the 'product' of the house and its location which households are consuming. The discourse of 'home' is explored and the meanings which households hold towards home and neighbourhood are examined.

The next section examines an appropriate research framework for empirical research on housing pathways. It is argued that a change of emphasis is needed from traditional, positivist research approaches towards techniques which capture the meanings held by actors and can adequately address dimensions of time. In particular it is argued that a research focus on interaction would help to elucidate both agency and structure elements of housing. Therefore this final section proposes a way forward for future research using the pathways framework.

The Social Constructionist Paradigm

Guba and Lincoln (1998: 200) define a paradigm as a 'basic set of beliefs that deals with ultimates or first principles. It represents a *worldview* that defines for its holder, the nature of the "world", the individual's place in it, and the range of possible relationships to that world and its parts ...' (emphasis in original). A paradigm has an appropriate ontology which defines the nature of 'reality', an epistemology which defines the nature of the relationship between the researcher and the world to be studied, and finally a methodology, which lays out the methods for finding out about the world.

As the editors make clear in their introduction, social constructionism is a broad church, with many different emphases within it. Therefore, it is necessary to define the particular approach taken before engaging in the main task of devising a conceptual armoury for the study of housing. The fundamental tenet of social constructionism is that reality is constructed by people through interaction. It is through interaction that people define themselves and the world they inhabit and so it is through interaction that the nature of the individual becomes apparent to themselves and to others. Language is a key element because it allows interaction to be detached from the 'here and now' and makes available to actors a vast accumulation of experiences and meanings. Language can be built up into zones of meaning or discourses that serve as a stock of knowledge which individuals use in everyday life and can be transmitted over time. Meaning is produced, reproduced, altered and transformed through language and discourse. Discourses are built into different 'sub-universes of meaning' which are held by different individuals or groups. The nature of the social order in a society will depend on the ability of people to be able to sustain their version of reality in competition with others. This raises issues about power relations.

Giddens' (1984) concept of structuration sits well within a social constructionist approach. Giddens argues that individuals act on the basis of their stored knowledge of what is appropriate in particular circumstances. In this way actors carry knowledge of social structures into interaction which, therefore, has both an agency and structural dimension. The outcome of interaction will depend on the power relations between the actors, part of which is the ability to sustain a particular knowledge or discourse in the minds of others. Clegg (1989) sees power as being manifest in the relationships between different actors who deploy different resources in 'games' played to a set of mutually accepted rules, which themselves are often challenged and in a state of flux.

An important element of power is the ability to achieve a desired outcome from a particular interaction, whether this reinforces or challenges existing social practices. This may depend on the relative resources of the agents to the interaction and their ability to use them effectively as well as the power relations around the particular structures.

Giddens places emphasis on the need to situate social practices in the context of time and space. Space is important because the locale of an interaction may influence its outcome. Time has many dimensions all of which may influence the nature of interactions. For example, the nature of any one interaction will be shaped by the previous experiences of the actors and of similar previous interactions. The experience of housing is dynamic, and for some people change may be very fast. An example is the homeless person who may change their housing situation almost every night. Clapham et al. (1993) draw the distinction between three different types of time in which housing pathways are structured: individual time (how old the individual is); family time (the stage of the family lifecycle); and historical time (what are the prevailing social, political and economic conditions). Different cohorts of people reaching retirement age, for example, will have experienced different circumstances which will influence their housing situation. Households forming in the 1930s would have had opportunities to enter owner-occupation not available in the 1920s.

This very brief exposition of a social constructionist approach forms the basis for the derivation of concepts to help in the analysis of housing. A number of key factors can be identified which need to be considered in the design of appropriate concepts. The subjectivist emphasis of the social constructionist approach places the meanings which actors hold at the centre of analysis. This is in contrast to most current housing research in the positivist paradigm which tends to objectify reality and ignore meaning. For example, households are assumed to perceive their circumstances in the same way as an outside observer and to hold simple and universal motivations.

The incorporation of structuration into the social constructionist approach means that there is a focus on the discourses which structure meaning and action. These are the zones of meaning which form the basis of what actors consider appropriate in particular circumstances. There may be many and conflicting discourses and a battle to achieve dominance between them. Discourses may be general, such as discourses of old age or youth which may influence attitudes and behaviour across a wide range of situations. Other discourses may be more specific to housing, such as the meaning of 'home ownership'.

Social constructionism also places emphasis on interaction. Therefore a conceptual framework needs to include a focus on the interaction between actors. Examples are the interaction between individuals within the household, between potential owner-occupiers and the staff of a mortgage provider, or between tenants and the housing officers employed by their landlord. Any conceptualisation of these interactions needs to incorporate an understanding of the nature of power and how it is deployed to influence the outcomes of interaction. Interaction has both agency and structural dimensions and a conceptual framework needs to be able to look at both and to examine the relationship between them.

These research issues of meaning, discourse, interaction, structure and agency, power, as well as the dimensions of time and space, are the factors which conceptual development within the social constructionist paradigm needs to address. The concept of a housing pathway is advanced as a way of meeting this challenge.

A housing pathway is defined as patterns of interaction (practices) concerning house and home, over time and space. This concept owes much to Giddens' idea of social practices and his borrowing of the time space geography of Hagerstrand. It is important to note that the concept of a housing pathway is the application of a metaphor to the field of housing. It is offered as a framework of analysis, that is, a way of framing thought. It is not a theory, although theories may be developed from its use either from empirical enquiry or analytical reasoning. Neither is it a research methodology, although it may provide a framework for one as we shall explore later. The concept of a housing pathway is now explored by focusing on four key elements.

Exploring Housing Pathways

1   The Household

The basic unit of a housing pathway is a household because this is the grouping in which people consume housing. The problem is that households are continually forming, changing and dissolving and so it can seem like building a conceptual framework on shifting sands. The processes of household change are key ones for any analysis of housing. For example, the creation of new households through young people leaving home will be an important influence on the number of houses required. The constituent structure of the household will have an impact on what people want from their housing in terms of size, location and design.

There are well documented changes to the form of households in Britain (for an overview see McRae, 1999). The number of households in England has increased substantially and is projected to increase further (from 19 million in 1991 to 21.9 million in 2011) despite slow overall population growth. The average household size is projected to decrease from 2.48 persons in 1991 to 2.27 in 2011. There is an increasing number of single person households and of lone parent families. Increasing rates of divorce are leading to more 'reconstituted families' where the links can be complex and the boundaries more permeable as links are maintained with first families. Also, people are living longer and there has been an increase in the number of older people living alone. These overall trends hide significant variations between ethnic groups and different regions of the country. Nevertheless, a number of general points can be made. The first is the increasing incidence of household types other than that of a married couple with children. Second, there is a growing individualisation of living arrangements with a growing proportion of adults living on their own or in households without another adult. Third, there is an increasing dynamic of change in households with people more likely to move through different household forms (for example, from single person to cohabiting couple to one-parent family to a reconstituted family).

The form which households take is shaped by both agency and structural elements. People choose their living arrangements in the light of practical issues (the housing available which they can afford) and the social mores of family, which are embedded in discourses which are used in political discourse and frame individual attitudes and expectations. Two main discourses can be identified which have dominated political debate in Britain namely the 'traditional family' and the 'diversity discourses'. Support for the 'traditional family' has come from a number of sources, including functionalist sociologists such as Parsons (1959). He argued that the traditional family of married couple with children fulfilled the functions of primary socialisation of children, inculcating them with the norms of society, and also gave its members emotional warmth and security to survive in a stressful world. More recently these ideas have been developed by writers of the 'New Right' associated with the governments of Margaret Thatcher in Britain and Ronald Reagan in the USA. Here the stress was also on the family as a bulwark against the encroachment of the state into the private domain. The alleged breakdown of the traditional family was blamed for the existence of an underclass which was disengaged from mainstream society and had led to increasing crime rates, drug abuse, educational failure and a dependency culture. The remedy for these problems was seen as strengthening the traditional family through public policy mechanisms designed to reinforce the male breadwinner role and encourage women to concentrate on the care of homes and families. Marriage could be encouraged through mechanisms such as council housing allocations policies which should discriminate in favour of traditional families. Morgan (1998) argues that government policy has provided too much support for families other than the two-parent norm.

An alternative discourse has emerged which places more emphasis on the diversity of forms family can take and accepts the validity of these forms. One element of the discourse is the feminist critique of the traditional family which is said to be based on an unequal relationship between men and women. The increasing diversity of the family is seen as being, in part, a consequence of the greater equality and financial independence of women. Rising rates of divorce are seen as partly the result of the increased ability of women to leave unsatisfactory or violent relationships because of their greater access to the labour market. The diversity discourse also draws on the wishes of some ethnic minorities who may have different cultural norms which may be based on arranged marriages and extended families. Gay and lesbian groups have campaigned for equal rights and social standing with traditional families to be able to adopt children or have legally recognised marriages.

The discourses of family are important for an analysis of housing pathways because they frame political debate and help shape policy mechanisms in housing and other public policy fields. They also frame the perceptions, attitudes and choices of people and households.

A cornerstone of the pathways approach is to seek to understand the meanings, perceptions and attitudes of households, as the household is the unit within which housing decisions are made. Therefore, understanding of how these decisions are arrived at and the factors which influence them is crucial to an understanding of housing outcomes. Most housing research has stopped at the front door and has analysed actions rather than seeking explanations for these actions. Alternatively generalised assumptions have been made about motivations which are rarely tested through empirical research. Where there has been a concern with the attitudes of households (see for example Cairncross et al., 1997) the focus has been on the views of individuals towards particular issues such as the service provided by their landlord. These views have not been seen in terms of their formation and sustenance within the household, or in the context of their relation to other views and their place in the holistic meanings held by households. The pathways approach seeks to discover and understand the factors which shape meanings and actions.

In an increasing number of instances the household is made up of one person who is, therefore the focus of research attention. However, in most cases the household is made up of more than one adult who would be expected to be part of a decision making process. Within the household a to some extent unique system of meanings and actions will be constructed from the negotiations and choices based upon personal and shared beliefs and histories (Dallos, 1997). Over time there is likely to emerge out of the interactions of members a set of rules or norms within the household over how to manage particular situations. For example, who chooses the internal décor or who has the final say over when and where to move and the factors that are to be taken into account. The process of interaction could be complex and will be influenced by power relations within the household as well as wider norms and values associated with family discourses.

Households may develop a long-term view of where they would like to be in the future and formulate a strategy to achieve this that will frame individual decisions. The existence of a strategy is a guide to the extent to which they engage in what Giddens calls life planning by actively seeking to organise and control their lives. McCrone (1994) and Anderson et al. (1994) found a high proportion of households in their samples that had housing strategies. However, in their study of teenage mothers, Allen and Bourke Dowling (1999) found that many were often following rather than controlling what was happening to them and felt powerless to influence the course their lives were taking. Little is known about the processes undertaken by households in making housing decisions as the next few sections will show. The pathways framework highlights the importance of filling this gap if we are to understand why people make the decisions which shape the nature of the housing experience.

2   Work

Employment is a key factor that underlies household structures and is an important factor in households' ability to afford housing. At the level of the individual household housing and employment pathways are often closely linked, with decisions in one sphere having important implications for the other. At the same time the structural discourses surrounding public policy in the employment field have had profound influence on housing policy.

The primary discourse in Britain has been that of a flexible labour market. This is both a description of current trends and an aim of government economic and employment policy. A flexible labour market is said to improve the productiveness of the economy by increasing the productivity of labour and allowing industries and individual companies to respond flexibly to changing tastes and demand. Flexibility is said to entail a growth in part-time work, flexible working hours or self-employment. People cannot assume that they have a job for life and may well have to retrain as technological change calls for different skills.

Housing has been viewed as a major constraint to the flexibility of labour. Rigidities, particularly in owner-occupation and public renting are said to reduce geographical mobility. Inefficiencies in housing production mean that housing shortages arise in areas such as the southeast of England where economic growth is strong. Resources in public rented housing have, in the past, often been directed at areas of poverty and substandard housing rather than at growth regions where housing conditions may be good but there is a shortage of affordable housing for key workers. The private rented sector, which is regarded as the most flexible tenure, is limited in size and geographical scope and has not responded as desired to government attempts to give it fresh impetus. All of these issues, which have arisen from the discourse of labour market flexibility, have been at the forefront of recent housing policy debate.

The perceived needs of the flexible labour market have set the agenda for housing policy in Britain in the last few years.

At the level of the individual household, decisions about housing and employment will often be inextricably linked. One of the major functions of a home is to provide access to employment opportunities and households may be faced with trade-offs between housing and employment goals. For example, the main earner in the household may want to take advantage of a job opportunity in another location. This situation could lead to a complex household decision-making process in which factors such as the benefits of the existing home and locational stability may be compared with the benefits of a new home and job. Individual household members may have different views and these will need to be reconciled through a decision making process in which members may have differential power. McCrone (1994) found in his research that the woman's employment and home situation was usually subordinate to the man's. However, there are few research studies that shed light on decision-making processes in different kinds of households. The pathways approach shows the value of this knowledge.

One study was carried out some time ago by Forrest and Murie (1987) who focused on a selection of affluent owner-occupiers in one area of Bristol. Their research technique was the biographical interview designed to illuminate what they called the housing histories of their sample. The people interviewed by Forrest and Murie were either self-employed, professionals in the public sector, executives working for expanding multinational companies, or small businessmen. A number were described as 'localist', having spent all or the vast majority of their housing history in the Bristol area, mainly because they were tied to family businesses in the area. However, most were very mobile having lived in a number of places because of career moves and they expected to move again in the near future. These moves were either with the one employer or involved changing employers, but they were usually supported by employers through financial and practical means. The high level of mobility of this group influenced their attitude towards the purchase of housing. They were generally concerned to maximise the equity value of their house by choosing dwellings considered to be highly saleable and by investing as much as they could afford. This was done not to generate wealth, as they were already well rewarded through their jobs, but to facilitate mobility. Many had thoughts of moving to high value areas such as London, because that is where employment opportunities were concentrated, and wanted to be able to do so without sacrificing too much in terms of housing quality.

Forrest and Murie conclude that the housing histories of the households in their sample were dominated by their employment circumstances. In particular, the male job was the driving force, even though a number of their wives were in employment. The husband's work came first and the housing situation was designed to support this. Breugel (1996) draws attention to the implications of the dominance in some households of the male career in determining migration decisions for what she calls the 'trailing wife'. She argues that trailing wives are as prevalent as ever despite the increasing number of women in employment. In many cases the market for 'women's work' is local and so many wives feel they can pick up equivalent employment in a new location.

For Murie and Forrest's movers, employment considerations determined their general location namely the city they lived in and so they were dependent on the geography of their industry. Nevertheless, the specific location or neighbourhood was strongly influenced by family considerations such as the availability of good schools for the children.

The fieldwork for the research was undertaken over 20 years ago and Forrest and Murie comment on the historical specificity of the experiences of the households. However, the research highlights a group of people for whom housing choices are strongly influenced by their employment circumstances and whose housing pathway was dominated by their employment pathway.

This study is an example of the kind of research which can illustrate the links between work and housing at the household level. However, the employment categories covered are very narrow and specific. Much more needs to be known about households in other employment circumstances and with different priorities between work and housing.

3   Paying for Housing

Complex institutional forms have been constructed to enable households to pay what can be a large proportion of their total expenditure for housing. The institutional forms tend to be tenure specific and are characterised by extensive government intervention. Government involvement is justified on the basis of political discourses and the example of globalisation is explored briefly here.

Globalisation has many dimensions (for a review see Waters 1995) but two are considered here. The first is the increasingly global reach of financial markets and the existence of what Waters (1995) calls 'floating finance' where money moves freely and rapidly between countries and uses. This has meant that the housing finance system in Britain has become more integrated into these global flows. Associated with this change has been a discourse which has endowed the concept of globalisation with neoliberal values and meanings of a consumerist free market world (Steger, 2003). The discourse has emphasised the inevitability of the free flows of capital, seeing globalisation as a natural force which governments are ill-advised to attempt to resist. Globalisation has become what Steger (2003) calls a 'strong discourse' which is 'notoriously difficult to resist and repel because it has on its side powerful social forces that have already preselected what counts as "real" and, therefore, shape the world accordingly' (Steger, 2003: 96). The globalisation discourse has dominated discussion and policy towards the finance industry since the 1980s when it underpinned a restructuring of housing finance for owner occupation (for a review see Clapham, 1996). Barriers to entry to the housing finance sector were reduced, resulting in high street banks competing with traditional building societies. Restrictions on building society activities were relaxed and societies were given the option of turning themselves into public limited companies along the lines of the existing high street banks. The previous practice of collusion on the setting of interest rates was outlawed and competition encouraged. The system of credit controls operated by the Bank of England was dismantled as part of a general deregulation of the financial sector. Government relied on the policy instrument of interest rate changes to regulate both the housing sector and the wider economy.

Although these changes have often been termed 'deregulation', government intervention still exists, albeit of a different kind. The emphasis of government regulation now seems to be to ensure barriers to competition are minimised and consumers are protected from unfair practices. An example of the latter is government action to insist on the provision of certain information to consumers, for example over the selling of endowment mortgages. The discourses surrounding policy for the private sector are ones of choice, efficiency and consumer protection.

The results of the changes are diverse and wide-ranging. Competition has resulted in a wider range of financial products available to consumers and has ended mortgage queues. However, in the new interest rate competition the traditional mutual building societies have consistently undercut their rivals the private banks. This has not stopped the rush of mutual societies to change their status or restricted the capture of large parts of the market by the banks. There is also a concern that the new system has resulted in increasing volatility of house prices (Stephens ,1995) which has a destabilising impact on the economy as a whole. The volatility of the housing sector is now seen as a major barrier to entry to the Euro zone and fiscal instruments are being sought in order to give government more leverage over market fluctuations.

The deregulation of the institutional structure for the finance of owner-occupation has created a framework in which households make their own housing decisions. This framework is characterised by continuing long-run real increases in house prices, but increasing short-term volatility. The problems of mortgage arrears and negative equity caused by increases in interest rates and the sharp decrease in house prices in the early 1990s have emphasised the risks associated with house purchase.

Little is known about households' perceptions of the options open to them and how it influences their behaviour. For example, it is known that people with similar income and family characteristics spend very different amounts on housing (Kempson, 1993) but not why this is so. Households also differ in the importance they attach to the wealth accumulation aspects of the ownership of a house. It is difficult to assess the importance of financial factors in the decision to become owner-occupiers. These decisions are made on a wide variety of grounds in which financial issues are intertwined with a large number of other considerations. The perception that owner-occupation is a secure investment has become an accepted part of the lexicon of 'home ownership' in that it provides a secure investment for the future or 'something to pass on to the children'. However, it is difficult to assess how important this factor is because of the many different attributes of tenure, such as house and neighbourhood quality, and rights and obligations of occupancy. It is likely that the relative importance of the investment factor relative to the others may vary between households.

It is just as difficult to disentangle investment issues from others in the decision to move house. Very few households say that they move to a more expensive house in order to build up more equity. Moves are usually made for reasons of changes in employment or family circumstances. However, households may take advantage of the build up of equity in order to 'move up the housing ladder' even if the main reason for moving is different. They may have in their minds a strategy of house moves that will reflect the ability to build up and reinvest equity and to take out new mortgages. This may be done in order to gain better quality housing, but it may be done in order to acquire wealth that will be released for spending on other areas. Very little is known about how different households perceive these situations and what their motives are.

It was noted earlier that households spend very different proportions of their income on housing. However, it is not clear whether this is related to different attitudes towards wealth acquisition. Do households concerned with building up wealth spend more on housing than households who are primarily motivated by consumption attributes?

The question of attitudes towards wealth acquisition is also important in decisions on house repair and improvement. To what extent are households motivated to spend money on their properties by concerns over whether the expenditure will add to the asset value of the house? Here again it may be difficult to disentangle the many factors involved. For example, the decision to add an extra room to a house (a conservatory or loft conversion) may be motivated by a mixture of concerns about current consumption (the increased pleasure from use of the extra space) as well as asset accumulation (will it add to the value or price of the house?). It may also be tied in with decisions about whether to improve the present house or to move to another. Again this may raise a mixture of financial and other considerations. The problem with all of these issues is that the same behaviour may be motivated by different perceptions and attitudes.

Some insight into the different ways that households can perceive and react to their circumstances is given by research on mortgage arrears. Despite the perilous position in which many households in mortgage arrears found themselves and the lack of control over their circumstances which many felt, they still actively pursued strategies to cope with their situation and to improve their position. Many continued a dialogue with their lenders, keeping them informed of the situation and attempting to negotiate reduced payments. This strategy sometimes worked for a limited period of time, but many lenders lost patience as time wore on. Households tried to find practical solutions to their difficulties in a wide range of ways. For example, Burrows and Nettleton (2000) give examples of households borrowing money from family members or friends. Some decided to use 'last resort' lenders who gave loans at high rates of interest which provided short-term relief, but compounded the financial problems in the medium term. Some households looked at ways of supplementing their income by working longer hours or taking in a lodger or tried to reduce their expenditures. If these strategies were unsuccessful and households lost their home, they were then faced with trying to find somewhere to live and mitigating the impact on important areas of their lives such as children's schooling. Finding somewhere to live often involved negotiations with local authority housing departments and other landlords. The loss of a home meant that household routines were often severely disrupted and had to be re-established in often difficult circumstances. Loss of the home was usually associated with a low income as well as possibly a large outstanding debt which lenders could insist on being repaid.

The change of lifestyle and categorical identity involved often challenged people's self-esteem and ontological identity and created scars which people carried for the rest of their lives.

In summary, little is known about the reasons for household behaviour in relation to payments for housing. Why do some people spend more on housing than others? How important are financial considerations in housing decisions? How do households plan their housing expenditures and how do they react to problems or changes in circumstances? The ability of people to pay for housing is determined by their income and wealth. In turn this is influenced by a wide variety of factors such as labour market conditions and discourses such as 'flexibility' which frame labour market behaviour. Government intervenes in the distribution of income and wealth through taxation and welfare benefits policies. These interventions are accompanied by discourses which frame them, such as globalisation which are related to other discourses such as labour market flexibility. The institutional structures of housing finance mediate between income and wealth and housing payments.

4   Home and Neighbourhood

In moving through their housing pathway households are consuming houses which hold meaning for them. This meaning is related to the physical structure and location of the house, but is not synonymous to it. In other words households will differ in the meaning they attach to the same physical surroundings. This is a major challenge to traditional approaches to housing policy which has focused on the physical structure and condition of housing on the assumption that this is an appropriate measure of quality and will ensure that households are satisfied with their housing. However, there is a lack of a consistent relationship between household views of the quality of houses and the defined standards. This was shown most clearly in the slum clearance campaigns of the 1970s when many houses were cleared when most residents were happy with them (Gower-Davies, 1972). Another example is the greater satisfaction that older people feel towards their housing despite living, on average, in worst conditions.

Houses have meaning to those who live in them and to others. For example, at the societal level, the design of houses can be symbolic reflections of a particular discourse. 'Homes fit for heroes to live in' were produced after the First World War on garden city lines for the 'aristocracy of labour' who the government were keen to win over as a bulwark against socialism. The housing design symbolised a commitment to social reform. The design of a house can influence the view held by passers-by of the status of people who live in it (Nasar, 1993). Purchasers of new housing are wooed by show houses which attempt to link the dwelling with a preferred lifestyle (Chapman, 1999).

The internal layout and use of a house reflects attitudes towards family life. For example, the increased differentiation of internal space is associated with a move from a discourse of 'collective family life' to one of 'individualised self-fulfilment' (Chapman, 1999). The position of the kitchen in the house has moved from being isolated at the back, to being an integral public room which serves as the main focus of the house. This reflects changes in the use of servants in the home and the changing social role and status of women (Craik, 1989).

There has been a considerable literature on the meaning of home which has been dominated by concerns about the impact of housing tenure (see Saunders, 1990). Also the emphasis has been on the construction of lists of the constituents of home with little attention paid to differences between people in the weighting they give to these elements. Far less importance has been attached to examination of the differences between individuals in their meaning of home and its relationship to other lifestyle factors.

A house derives meaning from its location and provides access to a wide range of facilities. There has been an increasing debate about the nature of neighbourhood and local social relations. The extent of local social interaction in the neighbourhood seems to be very different for different people. Some households where all the adults are working may interact with neighbours in the local area only rarely. Friendships may be made through work or hobbies and their lives outside the home may take place largely in locations away from the local neighbourhood. Others who are not in work or who may have mobility constraints or low income may be confined to the local area all day and may be dependent on neighbours for friendship and social contact. Forrest and Kearns (2001) argue that the quality of neighbouring is an important element of people's ability to cope in disadvantaged neighbourhoods. In more affluent areas, however 'neighbourhood may be rather more important then neighbouring – people may "buy into" neighbourhoods as physical environments rather than necessarily practice a great deal of interaction' (Forrest and Kearns, 2001: 2130). Scase (2000) argues that people still obtain a sense of attachment to the neighbourhoods in which they live even though they do not use the local facilities extensively. A key factor in this attachment is the 'brand' or image of the area. Butler and Robson (2001) found that differences in the image of neighbourhoods were reflected in the attitudes of residents. The link between identity, lifestyle and neighbourhood is a very interesting and important area of research which the pathways approach highlights.

The existing work on the meaning of home is important, but it has been treated as a discrete topic and not linked with the other aspects of the housing pathway outlined here as well as with other factors such as identity and lifestyle. The major conceptual task is to link together these dimensions of housing in order to further a holistic understanding of what housing means to households and its role in providing a means of personal fulfilment. The next section looks at ways in which this task can be taken forward.

Analysing Pathways

Sarre (1986) has laid out a useful framework for applying the concept of structuration to empirical research which is relevant to the task of elucidating housing pathways. He breaks the research task into four elements as follows:

1 the elucidation of frameworks of meaning using ethnographic or biographical methods to clarify individuals' knowledge of the social structure and their reasons for action;
2 investigation of the context and form of practical consciousness;
3 identification of the bounds of knowledgeability to discover the unacknowledged or unconscious meanings held by individuals and the unintended consequences of actions;
4 the specification of structural orders, i.e., the structural factors which impinge on actions.

In other words, the research needs to employ ethnographic or biographic methods to understand the meaning of individuals and households and the conscious aspects of behaviour. However the unconscious meanings and actions also need to be explored bearing in mind the constraints and opportunities which structure them and are reproduced by them. Also the structures themselves need to be analysed.

There are a number of elements of an analysis of pathways, but not all of these need to be included in any one empirical research study. Indeed it would be very difficult to design and implement research which did undertake all elements simultaneously. Therefore, concentration on some aspects of the whole is usually necessary, although it must be stressed that all of the elements need to be in place for a full understanding of pathways. One of the strengths of the approach is that it draws attention to the importance of a comprehensive analysis. The emphasis on meaning and the wide-ranging nature of many discourses means that the pathways approach tends towards holistic forms of understanding. Many positivist research studies look for partial explanations of phenomena by examining the influence of one or a small number of variables in what is called a nomothetic explanation. In contrast an idiographic approach attempts to develop as complete an explanation as possible. Therefore concepts need to be derived which are holistic, enabling all the factors which influence meaning and behaviour to be related together. Where research constraints mean that only a nomothetic analysis is possible, there needs to be a framework in place where partial pieces of the jigsaw can be related to the rest of the picture.

Research on housing pathways needs to be able to capture the meanings held by households and others, whilst incorporating an awareness of the importance of interaction and the dimension of time. It is a difficult task to choose research tools which meet all of these criteria. Many of the current widely used research tools in housing analysis are positivist, thus not being 'adequate at the level of meaning' (de Vaus, 2001: 11) and are cross-sectional in design, thus limited in their approach to the time dimension. In addition, the distance of most research tools from interaction means that this aspect has been relatively neglected.

As de Vaus (2001: 235) argues:

> actions have meanings to people performing those actions and this must form part of our understanding of the causes and meaning of any behaviour. To simply look at behaviour and *give* it a meaning rather than *take* the meaning of actors is to miss out on an important source of understanding of human behaviour (emphasis in original).

Some of the elements of the pathways approach build on traditional modes of research. For example, the meanings and perceptions of households can be elucidated using biographical or other forms of in-depth interviewing. Structures can be examined through discourse analysis. There has been an upsurge in the use of discourse analysis in the field of housing in the past few years (for a review see Hastings, 2000) and it has proved useful in elucidating meaning at the societal level. However, there has been little research on meaning at the level of the individual household and its relationship to wider discourses. Where such research has been undertaken, as for example in the literature on the meaning of home, the primary research technique employed has been the semi-structured interview. In this research tool households are encouraged to relate 'the story' of their lives, or at least the part of it of interest to the researcher. The semi-structured interview has a long pedigree in social science research and its strengths and weaknesses are well rehearsed. It can be an effective way of entering the assumptive world of the subject and thus capturing the meaning they attach to their situation.

It is a mistake to equate one particular research design such as the pathways approach with particular research methods. As Marsh (1982) argues, survey research has not traditionally been good at tapping into subjective meaning as meaning was usually imposed from the outside by the researchers. Nevertheless, this does not mean that subjective meaning cannot be incorporated into survey research. Structured surveys can be designed in a way that allows respondents to project their meaning. Surveys can be useful in allowing the generalisation of data, because one of the problems of in depth qualitative interviews is that it can be difficult to generalise reliably beyond the individual case. However, if qualitative surveys can be effectively linked with more quantitative survey approaches the level of meaning can be effectively considered and appropriate generalisations made.

A comprehensive way of examining the structural, conscious and unconscious elements of meanings and actions is through the analysis of interactions. These may be of one household with other households living in the neighbourhood or with professionals such as housing officers or building society staff. These interactions could involve different sets of meanings and involve the kind of power games highlighted earlier. For example, Clapham et al. (2000) examined the interactions between housing officers and tenants. These often involved a battle over meanings associated with the category of tenant or the rules of the game which framed interactions. Tenants would put forward their situation as they saw it and the housing officer would interpret this in the light of predetermined categories based on organisational policies and procedures. Implicit in these policies and procedures and the way they were implemented were conceptions of appropriate behaviour by 'good' tenants. Behaviour which accorded with this norm was rewarded and inappropriate behaviour punished by, for example, refusing to use discretion to favour the tenant. In these interactions the housing officer was in a powerful position, with the tenant being a supplicant and not always possessing the knowledge or the skills to be able to challenge the judgement of the housing officer. Such interactions structured the nature of the landlord tenant relationship.

Implicit in the social construction of the 'good' tenant – i.e., the categorical identity of tenant – were wider structures. These included the concept of social exclusion which defined the nature of poverty and framed the way poor tenants were seen. Conceptions of the appropriate role of council housing as a tenure and its role were also important. Ideas of appropriate lifestyles and behaviour as a neighbour, which had a social class dimension, were also implicit in the actions of some housing officers. Housing officers usually came from a different social class than the tenants. Therefore, an understanding of a housing pathway is dependent on analysis of the more structural social constructions which framed interactions and the meanings held by households and others. In turn these wider structures were partly reproduced through housing interactions. Research needs to be able to understand processes of interaction. This can be done by reconstructing interactions from interviews with participants or observers. However, a more direct and effective technique is to observe the interaction itself through forms of participant or non-participant observation.

Government housing policy is often important in mediating between wider social structures and household pathways. The analysis of policy should involve both levels of action and discourse. That is not just the description of policy mechanisms and the way they are implemented, but also the language and meanings of policy documents. The language of policy influences the meanings held by actors and therefore frames interactions.

Research needs to incorporate the important dimension of time. However, most empirical research is cross-sectional in design because of the practical constraints of research funding and administration which tend to militate against long-lasting research projects. As de Vaus (2001) points out, cross-sectional designs are unable to unambiguously establish the time sequence in which events occur. Because of this they face the problem of identifying causal factors and causal direction. Much housing research is content to examine differences between groups at one point in time. This approach neglects the flows of households between different situations over time. For example, research on homelessness has shown that the number of people sleeping rough at one point in time is a fraction of those who sleep rough at some time over a period (Fitzpatrick, 2002). An understanding of homelessness needs to embrace the dynamics of this movement into and out of different situations.

The semi-structured interview which is used in much research attempting to elucidate meaning is essentially a cross-sectional research tool. However it can be given a time dimension in a number of ways. One way is to encourage interviewees to talk about their past lives. This biographical interviewing has been criticised because of the way that interviewees may tell the story of their life by reconstructing their past through a mixture of selective memory and hindsight. Therefore, it is argued that it provides little insight into how they felt at the time as past events, attitudes and perceptions are filtered through the lens of current understanding. The very act of recounting the past in a story gives order to a diverse set of experiences. 'Thus a coherent and explanatory narrative is carved out of a set of diverse experiences and a set of past identities are assembled to account for a present identity' (Hockey and James 2003: 210).

Hockey and James (2003: 85–6) ask important questions of the biographical method:

> we might ask to what extent can life histories, enlisted by a researcher for a particular purpose, be regarded as unmediated and unambiguous accounts of the complexities through which social life has taken place in another geographical setting or epoch? How far does the translation from verbal account to written transform the meaning of a life recalled and what roles do memory and forgetfulness play?

Despite these problems biographical interviewing can be a useful tool which provides insight into the way that people construct their lives over time (Gearing and Dant, 1990).

A way of avoiding the pitfalls of retrospective interviewing is by interviewing the same households at different points in time. This can be done in the form of a panel survey where the same households are tracked over time. Panel surveys such as the British Social Attitudes Survey or the Child Development Study provide valuable information about change over time. However, in the field of housing they have not been used primarily to elucidate meaning as they have usually been based on structured questionnaire surveys. A panel survey, which took the exploration of the meaning of housing as its primary focus and used semi-structured interviews would be extremely valuable.

The emphasis so far has been on the pathways of individual households and the assumption made that research would be on understanding these individual pathways. The housing field is viewed as being made up of the sum of all of the individual pathways. The key question then is how to move from the level of the individual pathway to a broader level to enable generalisations to be made. This may be necessary in order to design national housing policy, or to characterise the housing system in comparison with other countries. In any analysis the basic unit is the pathway of the individual household, but it may be possible to discern general patterns and so to be able to construct generalised categories of pathways through deduction from empirical research. Some pathways are small tracks and others are motorways in the sense that the route is shared with many people. Clearly analysis of the motorways is a fruitful way of understanding the key features of housing in a country, although a way could be found to indicate the range or spread of experience. Generalised pathways could also be inductive, that is, constructed as ideal types based on theorising, which could then be used as hypotheses for empirical research. Typologies are of use in categorising households in a way which can be useful for understanding and policy making. Policy making needs typologies if it is to be based on an understanding of the reactions which different kinds of households will have to policy mechanisms. Coolen (2003) argues that data on meaning (what he calls less structured data) can be analysed in the same way as structured data using statistical techniques. The only difference is that the important function of categorisation is undertaken during the data-collection, data-processing and data-analysis phases rather than being undertaken before the collection of data, which is the common approach in much survey research. The challenge is to find ways of categorisation which retain the richness and holism of the original data.

In generalising from individual pathways there are a number of factors which must be retained in any generalisation. The first is the focus on meanings. Although the precise meaning of their housing held by a household will vary because of their own particular situation, there may be some features which are held in common by a number of households. Existing work on the meaning of home has succeeded in achieving the balance between making generalisations whilst retaining the importance of subjectivity. The second is the characterisation of households as creative agents acting upon, negotiating and developing their own housing experience through life planning and lifestyle choice.

The third crucial element is the dynamic nature of pathways. The importance of change over time must be retained if the insights generated by the approach are not to be lost. The fourth element is the importance of social practices and the factors which frame them. Particularly important here is what has been called the politics of identity and the resultant categories and their associated discourses which frame public policy and the expectation and attitudes of households.

Conclusion

This chapter has attempted to outline the pathways approach as a way of providing a bridge between the general paradigm of social constructionism and its empirical application in the field of housing. The concept of structuration was needed to show the links between agency and structure and a concept of power employed to offer a framework for understanding differences in the constraints and opportunities facing households and their differential ability to make their life.

The pathways approach places the meanings of a household at the forefront of the analysis of their housing experience, but places this within the structural discourses which frame perceptions and attitudes. It focuses on the interactions between households and other actors in housing which embody both agency and structural dimensions and serve to restructure or to change structural discourses.

The pathways approach was explored through a review of important dimensions of a housing pathway such as the nature of the household and family and the links between housing and work and the impact on the differential ability to pay for housing. The meaning of home and neighbourhood were also explored. This review stressed the need for further conceptual and empirical development of the concepts of identity and lifestyle and their links to the meaning of home and neighbourhood.

Finally, the implications of the adoption of the pathways approach for research methods in housing have been outlined. A research focus on interaction has been advocated as a way of conducting holistic research which captures both agency and structural dimensions.

The pathways approach promises a fresh look at housing by building on the strengths of the social constructionist tradition of research which has concentrated on issues of meaning, by incorporating this work within a coherent and holistic analytical framework. There is clearly much further work to do in elucidating key concepts in this framework and applying them to the study of housing. For example, the concepts of identity and lifestyle have been used loosely here without defining them precisely or relating them together in a rigorous way. These are key concepts if we are to understand the role of housing in furthering personal fulfilment. As well as conceptual development, innovation is needed in techniques of data collection and analysis. In particular, the ability to generalise by constructing useful categorisations from unstructured data is crucial to the successful application of the pathways approach.

This brief review has also drawn attention to the many gaps in our current empirical knowledge of housing which the pathways approach throws into sharp relief. In particular the lack of understanding of the dynamics of household planning and decision making on key aspects of housing as part of the making of lifestyle choices is an important gap to fill. This task will demand a different approach to housing research, with more emphasis on understanding the interactions which shape the housing experience of households.

References

Allen, I. and Bourke Dowling, S. (1999), 'Teenage Mothers: Decisions and Outcomes', in S. McRae (ed.), *Changing Britain: Families and Households in the 1990s*, Oxford: Oxford University Press: 334–53.

Anderson, M., Bechhofer, F. and Kendrick, S. (1994), 'Individual and Household Strategies', in M. Anderson, F. Bechhofer and J. Gershuny (eds), *The Social and Political Economy of the Household*, Oxford: Oxford University Press: 19–67.

Breugel, I. (1996), 'The Trailing Wife: A Declining Breed?', in R. Crompton, D. Gallie and K. Purcell (eds), *Changing Forms of Employment*, London: Routledge: 235–58.

Burrows, R. and Nettleton, S. (2000), 'What Role for Housing Studies in the New Paradigm for Welfare Research? A Case Study of Families Experiencing Mortgage Possession', paper given to the Housing Studies Association Conference, University of York.

Butler, T. and Robson, G. (2001), 'Social Capital, Gentrification and Neighbourhood Change in London: A Comparison of Three South London Neighbourhoods', *Urban Studies*, Vol. 38, No. 12: 2145–62.

Cairncross, L., Clapham, D. and Goodlad, R. (1997), *Housing Management, Consumers and Citizens*, London: Routledge.

Chapman, T. (1999), 'Stage Sets for Ideal Lives: Images of Home in Contemporary Show Homes', in T. Chapman and J. Hockey (eds), *Ideal Homes? Social Change and Domestic Life*, London: Routledge: 44–58.

Clapham, D. (1996), 'Housing and the Economy: Broadening Comparative Housing Research', *Urban Studies*, No. 33: 631–47.

Clapham, D. (2002), 'Housing Pathways: A Postmodern Analytical Framework', *Housing Theory and Society*, Vol. 19, No. 2: 57–68.

Clapham, D., Franklin, B. and Saugeres, L. (2000), 'Housing Management: The Social Construction of an Occupational Role', *Housing, Theory and Society*, Vol. 17, No. 2: 68–82.

Clapham, D., Means, R. and Munro, M. (1993), 'Housing, the Life Course and Older People', in S. Arber and M. Evandrou (eds), *Ageing, Independence and the Life Course*, London: Jessica Kingsley: 132–48.

Clegg, S. (1989), *Frameworks of Power*, London: Sage.

Coolen, H. (2003), 'The Measurement and Analysis of Less Structured Data', paper to the International Conference *Methodologies in Housing Research*, Stockholm, Sweden, 22–24 September.

Craik, J. (1989), 'The Making of Mother: The Role of the Kitchen in the Home', in G. Allan and G. Crow (eds), *Home and Family: Creating the Domestic Sphere*, Basingstoke: Macmillan: 48–65.

Dallos, R. (1997), 'Constructing Family Life: Family Belief Systems', in J. Muncie, M. Wetherell, M. Langan, R. Dallos and A. Cochrane (eds), *Understanding the Family*, 2nd edn, London: Sage: 173–211.

De Vaus, D. (2001), *Research Design in Social Research*, London: Sage.

Fitzpatrick, S. (1999), *Young Homeless People*, Basingstoke: Macmillan.

Forrest, R. and Murie, A. (1987), 'The Affluent Home Owner: Labour Market Position and the Shaping of Housing Histories', in N. Thrift and P. Williams (eds), *Class and Space*, London: Routledge and Kegan Paul: 330–59.

Forrest, R. and Kearns, A. (2001), 'Social Cohesion, Social Capital and the Neighbourhood', *Urban Studies*, Vol. 38, No. 12: 2125–43.

Gearing, B. and Dant, T. (1990), 'Doing Biographical Research', in S. Peace (ed.), *Researching Social Gerontology: Concepts, Methods and Issues*, London: Sage: 143–59.

Giddens, A. (1984), *The Constitution of Society*, Cambridge: Polity Press.

Gower-Davies, J. (1972), *The Evangelistic Bureaucrat*, London: Tavistock.

Guba, E. and Lincoln, Y. (1998), 'Competing Paradigms in Qualitative Research', in N. Denzin and Y. Lincoln (eds), *The Landscape of Qualitative Research*, Thousand Oaks: Sage: 195–220.

Hastings, A. (2000), 'Discourse Analysis: What Does it Offer Housing Studies?', *Housing, Theory and Society*, Vol. 17, No. 3: 131–8.

Hockey, J. and James, A. (2003), *Social Identities Across the Lifecourse*, Basingstoke: Palgrave Macmillan.

Kempson, E. (1993), *Household Budgets and Housing Costs*, London: Policy Studies Institute.

McCrone, D. (1994), 'Getting By and Making Out in Kirkcaldy', in M. Anderson, F. Bechhofer, and J. Gershuny (eds), *The Social and Political Economy of the Household*, Oxford: Oxford University Press: 68–99.

McRae, S. (ed.) (1999), *Changing Britain: Families and Households in the 1990s*, Oxford: Oxford University Press.

Marsh, C. (1982), *The Survey Method: The Contribution of Surveys to Sociological Explanation*, London: George Allen and Unwin.

Morgan, P. (1998), 'An Endangered Species', in M. David (ed.), *The Fragmenting Family: Does it Matter?*, London: Institute for Economic Affairs: 68–99.

Nasar, J. (1993), 'Connotative Meanings of House Styles', in E. Arias (ed.), *The Meaning and Use of Home*, Aldershot: Avebury: 143–68.

Parsons, T. (1959), 'The Social Structure of the Family', in R. Anshen (ed.), *The Family: Its Functions and Destiny*, New York: Harper and Row: 241–74.

Sarre, P. (1986), 'Choice and Constraint in Ethnic Minority Housing: A Structurationist View', *Housing Studies*, No. 1: 71–86.

Saunders, P. (1990), *A Nation of Home Owners*, London: Unwin Hyman.

Scase, R. (1999), *Britain Towards 2010: The Changing Business Environment*, London: Department of Trade and Industry.

Steger, M. (2003), *Globalization*, Oxford: University Press.

Stephens, M. (1995), 'Monetary Policy and House Price Volatility in Western Europe', *Housing Studies*, Vol. 10: 551–64.

Waters, M. (1995), *Globalization*, London: Routledge.

Chapter 7

Necessary Welfare Measure or Policy Failure: Media Reports of Public Housing in Sydney in the 1990s

Kathleen J. Mee

Introduction

Why examine the way housing problems are presented in the media? There are compelling reasons for such analysis. Both media studies analysts and those investigating policy have argued that journalists are key actors in policy development (Sabatier, 1993; Williams, 1999). The presentation of policy problems involves the creation of narratives about policy change that evoke causal stories (Stone, 1989). Causal stories are sensitive to media portrayals (Stone, 1989). The definition of policy problems in part involves political actors competing with each other to have their definition of a problem become predominant. The media is one arena in which this definitional battle takes place (Miller, 1993). As part of these definitional battles the media 'regularly simplify, sharpen and politicise an issue by attributing responsibility or blame' (Papadakis and Grant, 2001: 297). Political actors such as governments, oppositions and welfare organizations are acutely aware of the role of the media in presenting policy problems and actively attempt to intervene in media coverage. Thus newspaper articles both reflect the nature of on-going policy debates, and 'draw from existing narratives of social reality' (Jacobs, 2001: 133) provided by sources and journalists that frame popular understanding of events and policies.

This chapter brings together two strands of the tradition of social constructionism as it has been used in housing research: discourse analysis and analysis of social problems and policy narratives (Jacobs, Kemeny and Manzi, this issue, Introduction) by presenting an account of newspaper reporting of public housing by the *Sydney Morning Herald* in New South Wales (NSW) during the 1990s. This period was a crucial time for the development of public housing policy in Australia. At the beginning of the 1990s public housing was portrayed in the *Sydney Morning Herald* as a problematic tenure, but one that provided an important welfare measure for tenants. By the end of the decade the problematic and residual nature of the tenure was at the forefront of reports and the necessity of providing public housing was increasingly questioned. This discursive shift was a direct product of attempts by successive federal governments to sell a new solution to problems of affordability at the lower end of the rental market, in the form of Commonwealth Rent Assistance. The marketing of the solution of Commonwealth Rent Assistance involved presenting public housing as a failed intervention in the housing market. In particular the representation of public housing estates as dysfunctional was a crucial rhetorical strategy used to justify the policy change. The reporting of these events in the *Sydney Morning Herald* was uneven. In constructing their stories journalists engaged in debates with each other, as well as the government, concerning the nature of the policy changes enacted throughout the decade. The way that news stories are socially constructed to produce a range of narratives about housing change has not been explored adequately in previous housing literature. The next section outlines the role of public housing in NSW.

Background

By the year 2000 there were 130,000 public housing dwellings in New South Wales (NSW), with over a third of these being located on large estates containing 100 or more dwellings. Thirty-seven per cent of allocations to public housing were priority allocations, essentially given to those people in acute housing need. Most of the priority allocations were located on large estates (primarily because of the greater availability of stock in these locations). Only 10 per cent of public housing tenants paid full market rent, indicating their low levels of income and need for rental support (and in stark contrast to the mid-1970s when 60 per cent of tenants paid full market rent). The majority of tenants were welfare recipients, and many required other support services. In 2000 there were 97,000 applicants on the waiting list for public housing in NSW, a 50 per cent increase since 1990–91 (Directions, 1999).[1]

The 1990s, as Dodson (2001) notes, was a key period where reform of public housing provision was considered, debated and contested by various groups. The federal government was involved in developing policies, strategies and reports that addressed housing through much of the 1990s. Such activities included the National Housing Strategy papers (1990–92) and an Industry Commission inquiry into public housing that reported in November of 1993.

In 1995 the federal government released a policy document titled *Community and Nation*, which discussed problems associated with the provision of public housing. The election of the conservative Liberal-National Party Coalition in March of 1996 led to a series of developments, loosely termed by the government 'Commonwealth Housing Reform'. In 1997 the Federal Senate's Community Affairs References Committee held another enquiry into the nature and adequacy of housing assistance. These multiple investigations into housing provision provided a range of groups with the opportunity to intervene into debates about housing policy (Dodson, 2001).

The nature of housing provision was closely contested between the federal and state levels of government during the 1990s. This contestation was an important element in media coverage. As is the case with many forms of service provision in Australia, funding for public housing is provided by the federal government (through the Commonwealth State Housing Agreement or CSHA) but is administered through state governments. Debates around the provision of funding for public housing are therefore in part, debates between the federal government and the state governments about levels of funding. A central concern of debates during the 1990s was the mix of federal funds flowing to public housing through the CSHA and those funds flowing to poor renters in the private market through Commonwealth Rent Assistance (CRA). During the decade CRA increased in importance in real terms while federal government public housing funding declined in real terms (Caulfield, 2000). Politicians used debates around this funding mix to attack their political opposition in power at a different level of government. It is important to note, therefore, that in NSW during the 1990s the political party in power at the federal level was from a different political party from the government in power at a state level in all but one year (from March 1995 to March 1996).[2] The next section provides the framework through which the media articles about public housing were analysed.

Theory

Policy Problems, Housing Problems, Media Problems

This chapter is part of a growing body of work investigating the role of language and discourse in constructing housing and urban policy problems (Jacobs, Kemeny and Manzi, this issue, Introduction; see for example, Atkinson, 2000; Goodchild and Cole, 2001; Jacobs and Manzi, 1996; Jacobs,

Kemeny and Manzi, 2003a, 2003b; Marston, 2000; 2002). However, there is much further scope for examining the ways in which the characteristics of housing markets become interpreted as 'problems', and more specifically how the conceptualisation of these problems changes over time. Understanding how issues becomes problems is vital because, as Oatley notes:

> The framing of the problem, the language or discourse that is used is very important. It determines the way in which the problem is viewed, the causes that are thought to be operative, and the policies that are thought to be appropriate (Oatley, 2000: 89).

This chapter draws on the insights of the work on problem definitions in the following ways. Firstly, problem definition is in part, as Stone (1989) notes a process of 'image making'. Those defining problems attempt to create deliberate images that shape the understanding of the problem. Secondly, these image creations involve 'attributing cause, blame and responsibility' (Stone, 1989: 282). Thirdly, in constructing stories about policy problems 'political actors use narrative story lines and symbolic devices to manipulate so-called issue characteristics, all the while making them seem as though they are simply describing the facts' (Stone, 1989: 282). This process of 'simply describing the facts' legitimates the interpretations of political actors. Often, as Goodchild and Cole (2001) point out, such a strategy depends on the mobilisation of 'common sense' in understanding problems (p. 106). Thus the preferred understanding is presented as simple description, and the only sensible interpretation of the events or problems discussed. Fourthly, the way that a problem is defined, and who is deemed responsible for the problem is intimately linked to proposed solutions for the problem (Atkinson, 2000; Marston, 2002; Jacobs, Kemeny and Manzi, 2003a). Fifthly, the construction of problems and their policy solutions are contested (Hastings, 1998). Policy development therefore involves debates between groups seeking to ensure their own constructions of problems become predominant. An understanding of media reports therefore plays an important role in illuminating how policy debates are played out in the public realm (Jacobs, 2001).

Policy debates generally, and debates about welfare in particular, are an important part of 'routine news' (Putnis, 2001: 74). Much of the focus of authors examining media coverage of welfare issues centres on the negativity of media reporting of welfare programmes (see for example, Jacobs, 2001; Mieweld, 2001; Marston, 2002; Putnis, 2001). It is useful then to consider how journalists construct their stories and whether these constructions have a tendency towards negative portrayals of welfare.

Following the ground-breaking work of Entman (1991, 1993) media studies analysts have employed the concept of frames to explain the construction of media reports. Entman argues that employing frames involves two important elements, selection and salience:

> To frame is to *select some aspects of a perceived reality and make them more salient in communicating text, in such a way as to promote a particular problem definition, casual interpretation, moral evaluation, and or treatment recommendation* (Entman, 1993: 52, emphasis in original).

The media studies analyst Altheide (1997) has taken this analysis of frames a step further to argue that certain media articles are organised through a problem frame. Problem frames, Altheide argues, 'are part of a format organised around a narrative that begins with a general conclusion that "something is wrong" and we know what it is' (Altheide, 1997: 654). Hence the media may present issues in a way that is consistent with how political actors attempt to portray policy problems.

In presenting policy problems the media often draw on specific, stereotypical representations of places to make concrete their arguments (Burgess, 1985; Parisi and Holcomb, 1994). In an article about the depiction of lone mothers Duncan, Edwards and Song (1999) argue 'The targeting of particular public housing estates with a supposedly high proportion of young "underclass" single mothers seems to be a virtual social science technique for delivering messages' (241). Depictions of particular places as dysfunctional are particularly important to articles utilising a problem frame (Altheide, 1997). Journalists draw on 'common sense' understandings of places thought to exemplify policy failures in order to increase the salience of proposed policy solutions (Duncan, Edwards and Song, 1999).

How then can we understand the relationship between journalists who construct newspaper texts through a problem frame and political actors who are attempting to present their construction of policy problems and solutions? The suggestion of some media studies analysts is that the construction of media texts should be understood within the context of journalists working in an interpretive community. Berkowitz and TerKeurst (1999) define an interpretive community thus: 'a cultural site where meanings are constructed, shared and reconstructed by members of social groups' (125). The creation of news stories by journalists leads them to rely on sources, particularly institutional sources, for 'information subsidies' (Berkowitz and TerKeurst, 1999; see also Brindle, 1999; Franklin, 1999; Entwhistle and Sheddon, 1999; Putnis, 2001; Schudson,

1996). Media articles reflect both the framing actions of journalists and the sources that journalists draw upon in creating their articles. Sources are active participants in this process. Governments have been taking an increasingly proactive role in 'news management' (Papadakis and Grant 2001: 295; see also Berkowitz and TerKeurst, 1999; Franklin 1999; Gaber 2000) and welfare lobby groups also attempt to use the media to further their portrayal of events (Papadakis and Grant, 2001; Putnis, 2001).

Journalists are primarily responsible for deciding whose narratives will be selected and sources aware of this will package their stories through problem frames that the media can identify (Altheide, 1997). Journalists then become more dependent on the framing of the problem provided by the source (Entman, 1991). Sources, however, have various levels of skill and resources in getting their message across (Deacon, 1999). Government departments are often relatively well resourced and connected in an interpretive community compared to welfare or community activists for example. However even official sources will be constrained in how they present their arguments. Official sources from governments may be internally divided over the nature of a problem (Miller, 1993), or may be engaged in battles with other levels of government about the nature of a problem and who bears responsibility for the problem. Moreover, there will not necessarily be consensus between what official sources want to say and 'news values'. Brindle (1999), for example, notes that it is much more difficult to get 'good news' stories published than 'bad news' stories because of the way that journalists construct the news.

Other factors that influence the social construction of newspaper articles relate to the perceptions of journalists about the nature of journalism (Berkowitz and TerKeurst, 1999). Journalists may see themselves as the watchdogs of society (Berkowitz and TerKeurst, 1999), with a role in highlighting the inadequacies of government policies (Parisi and Holcomb, 1994), particularly in relation to the provision of services for the poor. Despite their reliance on official government sources they may therefore chose to write articles that are critical of government policy. Journalists also interact with other journalist colleagues to learn the expectations of their job and newspaper (see also Zelizer, 1993a, 1993b) and editorial regard for target markets (Papadakis and Grant, 2001: 297). Through these interactions journalists also learn common news narratives, which indicate how a story should be told and who should be involved in telling the story. Despite this, a journalists' ideological predispositions can still play an important role in determining what sort of articles they will produce (Papadakis and Grant, 2001: 297). The conflicting nature of such predispositions between journalists, and their differing role within the newspaper, may lead to disputes between journalists who are part of an interpretive community (Lindlof, 2002).

This chapter then looks at hundreds of individually created texts and their relationship to each other. Each particular newspaper report has its own individual history. The creation of these texts, their social construction, needs to be taken into account when interpreting media stories.

Methods

In order to examine how events in public housing were reported in the *Sydney Morning Herald* during the 1990s a collection of all articles referring to public housing from across the decade was assembled. Sabatier (1987) and Driedger and Eyles (2003) argue that in order to properly understand changing discourses that surround a policy debate a period of at least ten years is required. Other researchers who have drawn on media articles to examine housing (for example Forte, 2002; Marston, 2002) have concentrated on crisis moments in the depiction of a housing issue. However, as Myers et al. (1996) have pointed out, periods of crisis tend to generate simplistic and extreme examples of a discursive construction. By examining a broader sweep of articles, these crisis moments can be better understood and contextualised.

In order to successfully pursue media analysis in the tradition of critical discourse analysis, 'a comprehensive data base that is easily accessible' is required (Altheide, 2000: 293). The *Sydney Morning Herald*, the daily broadsheet newspaper in Sydney was chosen for the study due to its availability. The articles used for this research were collected via two search engines, the *Dow Jones Index* and the *Infoquick*[3] search engine. Articles in the news section of the newspaper were searched for over 20 key words or word combinations, such as 'public housing', 'Department of Housing', and 'Housing Minister'. Articles that were obviously irrelevant to the study were immediately excluded. After an initial cull, 2,213 articles were selected as being potentially relevant. All these articles were then read, and only those articles that pertained directly to public housing in New South Wales, or Australia more generally (but not the provision practices of other states) were included in the study. This provided a data set of 399 articles.

The articles were analysed using a procedure known as 'soaking' (Myers et al., 1996).[4] Soaking involves the combination of an extensive content analysis and intensive discourse analysis of key articles. Prior to content analysis all articles were studied, identifying themes and sub-themes within the articles.

Each article was then classified by year, theme, sub-theme, article type and journalist.

During the second stage an intensive discourse analysis was undertaken. This discourse analysis was conducted using a methodology developed by Mee (2000) drawing on the criteria in Table 7.1.

Table 7.1   Elements examined in discourse analysis developed by Mee (2000)

Characteristic	Importance in discourse
Word choice	What words are present or absent? How are words linked and ideas organised?
Accent	What is the stated context of the article? How are speaking positions indicated within the article? How do these affect the meaning conveyed?
Genre	What type of article is being examined: is it a news report, a longer exploratory essay or an editorial comment? How do these affect interpretation?
Underlying meanings and values	What are the explicit and implicit values that underlie this story? How do they appear?
Positioning of actors	What sources are drawn upon in the article? How are they given authority in newspaper accounts? What positions are marginalised or absent?
Attempts of groups to gain access to the media	What social groups are apparent in newspaper accounts? This is a further aspect of positioning.
Disarticulate or rearticulate	How are news events linked to other events? How are places explicitly or implicitly part of this process?
Variability, instability and stability	How stable or variable are the interpretations of events by the media? How do new meanings arise?
Intertextuality	How does this text relate to, or refer to other texts? What are the presuppositions of the article? What shared values are assumed?
Headlines	How do headlines direct readers to read the stories that follow in particular ways?
Visual images	How do visual images direct a particular reading of the article? Do visual images alter the accent of the article?

Note: This table was developed drawing principally on van Dijk (1988); Fowler (1991); Fairclough (1995); Pfau (1995); Altheide (1996); Altheide (2000).

Three features are particularly important to the discussion of the discourse analysis below. First, the role of underlying meanings and values in constructing the news stories. As Papadakis and Grant (2001) note articles that appear on the surface to be critical of government policies, may contain implicit messages about the importance of state welfare. Second, the position of actors in the articles is vital. Each article was examined for the sources mentioned and how they were drawn into the article (see Table 7.2 below). Third, newspaper articles need to be considered in their social and political context (Altheide, 2000; Jacobs, Kemeny and Manzi, 2003b). In order to understand this context the author drew on a number of texts to interpret the newspaper articles, including reports and press releases about public housing produced by the Commonwealth and NSW governments and housing lobby groups such as ACOSS and National Shelter.

Table 7.2   Named sources in *Sydney Morning Herald* articles concerning public housing 1990–99

Named sources	% of articles with this source
State government	33
No named source	25
Community, welfare or religious organisation	19
Local resident or service user	16
Federal government	10
Developer or industry lobby	6
Academics	6
Local community worker	6
Local government	5
Professional organisation or professional	5
Local business	2
Union group	1

Note: The percentages in this table do not add up to 100 per cent as any article may contain sources from more than one of the groups listed here.

Public Housing Reporting in the *Sydney Morning Herald*

Academic studies of the reporting of welfare have emphasised the negativity that characterises mainstream media accounts (Forte, 2002; Mieweld, 2002;

Putnis, 2001). Much more support for the provision of public housing was displayed in the pages of the *Herald* than would be expected, especially in the early years of the decade. However, the stigmatisation that accompanied discussions of public housing places provided a background for understanding public housing as 'failed'. By the end of the 1990s an understanding of public housing as a policy of the past was increasingly apparent in media coverage. I begin this discussion of reporting of public housing by examining the overall level of coverage of public housing issues. From there I discuss the major themes of public housing coverage, and focus on two key sub-themes of the coverage; the waiting list for public housing and the political commitment of governments to public housing provision. Drawing on this discussion of the overall nature of the coverage I move on to a more detailed analysis of the discursive shift in public housing reporting by focusing on articles from the early period of reporting, 1990, the middle period of reporting where ideas about public housing were most obviously being contested during 1995–96, and then drawing on articles from the end of the decade in 1999.

Public housing was a much more important news item at the beginning of the 1990s than at the end of the decade. As Figure 7.1 shows, the total number of articles on public housing was greatest in the first three years of the decade, peaked again in 1995 and then declined through to the end of the 1990s.

Why did the overall level of coverage of public housing decline over this period? The activities of government provide part of the answer. The beginning of the decade was characterised by a considerable amount of activity from the

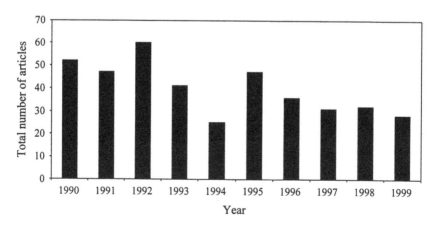

Figure 7.1   Total number of articles on public housing in the *Sydney Morning Herald* 1990–99

NSW Housing Minister in attracting media coverage to issues around public housing. However there is no simple relationship between levels of reporting and levels of ministerial activity in producing press releases, with 1996 producing fewer articles concerning public housing than the earlier years of the decade despite considerable press release activity from the minister's office. During the early period of the decade the activities of the federal government in commissioning reports for the National Housing Strategy and an Industry Commission inquiry pointed to public housing as a significant news issue.

A fundamental explanation for changes in the numbers of articles about public housing in the *Sydney Morning Herald* is that public housing became a much less important news story during the decade. Figure 7.2 shows the total number of articles published each year divided into four categories, news stories, feature articles, opinion articles and comment articles. Public housing as a news story declined significantly over the course of the decade. This is consistent with cycles of reporting on welfare issues in general (Papadakis and Grant, 2001), where stories concerning vulnerable groups peak and then diminish in public attention. In explaining the decline further we need to turn to the changing ways that the *Herald* reported particular themes during this period, and the implicit message that framed the articles presented. The articles were divided into four themes each with a number of sub-themes as is shown in Table 7.3. Any particular article could contain elements from a number of these themes.

Articles concerning the social characteristics of residents and their neighbourhoods comprised the largest group. These articles were concerned

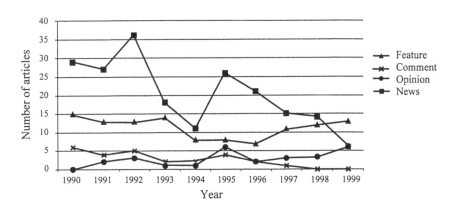

**Figure 7.2 Public housing article types, *Sydney Morning Herald*
1990–99**

Table 7.3   Public housing articles by theme and major sub-theme in the *Sydney Morning Herald* 1990–99

Theme	Major sub-theme[1]	Number of articles
Social characteristics of residents or neighbourhoods		216
Crime	72	
Unemployment, single parents, poverty	39	
Drug & alcohol use	22	
Health	22	
Child neglect	11	
Boredom	9	
Migrants or Aboriginal people	9	
Social problems	9	
Youth	9	
Neighbourhood disputes	5	
Nature of public housing provision		136
Waiting list	89	
Lack of facilities and services in neighbourhoods	19	
Poor or dangerous condition of dwellings	16	
Tenants rorts[2]	11	
General public housing policy		176
Political commitment of government to public housing	53	
Inner city public housing	21	
Home Fund crisis and public housing provision	19	
Public housing building approvals	14	
Department of Housing Management	11	
Estate Worker Programme	9	
Goods and Service Tax (GST) impacts	8	
Public housing spending	5	
Explicitly positive articles		36
Past	13	
Cost effective and stable	7	

Notes

1   Only sub-themes with more than five articles have been included in the table.
2   A rort is when a person takes unfair (not necessarily illegal) advantage of a system for their own benefit.

with criminal activity in public housing neighbourhoods, or the social disadvantages faced by public housing tenants individually or in groups. Such reporting on public housing is unsurprising, and is generally consistent with accounts of the pathologising of welfare provision (see for example Mieweld, 2002). These articles are entirely consistent with the use of a problem frame as outlined by Altheide (1997). There are different ways that the 'blame' for these social problems was apportioned in these articles. In some cases the articles clearly attributed the blame for social problems to the residents themselves (as Marston, 2002 argues was the case in his analysis of media reports in Queensland). Other articles however blamed the Department of Housing for the concentration of social problems found in public housing neighbourhoods, either because the department's allocation policy was seen as locating too many 'problem' tenants into certain locations or because of historic management failures (in keeping with the British analysis of Jones and Ward, 2002). In particular the building of large public housing estates was seen as responsible for concentrating and exacerbating social problems. The solutions proposed to the problems identified by articles in this category varied over the decade. Earlier in the decade government intervention, directly in public housing or in other service provision was implied. Later in the decade however the replacement of the supposedly failing estates with rent assistance in the private market was the implicit message of articles in this category.

The dominant category of articles concerning the nature of provision of public housing concerned the waiting list for public housing. The articles about the waiting list reveal interesting characteristics about the changing nature of reporting. As Figure 7.3 shows, there were more articles written about the waiting list for public housing early in the 1990s than by the end of the decade. As the waiting list for public housing grew over the course of the 1990s the absolute length of the waiting list does not explain this change. One reason for the changes in the coverage was that the Minister for Housing in the early 1990s drew specific attention to the waiting list through a number of press releases in 1990 and 1992. This was one instance where the relationship between the press releases and the press coverage was not a straightforward one however. While Minister Joe Schipp's press releases stressed the advances made by his department in relation to the waiting list the media coverage emphasised the problems of people waiting for extended periods for public housing.

The coverage of the waiting list altered more during the 1990s than is indicated by the total number of articles published on the issue. In the early period, articles written about the waiting list were considerably longer, and drew on stories of the problems of people on the waiting list, their poor housing

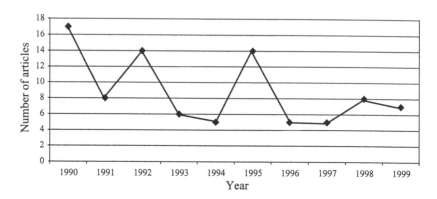

Figure 7.3   Total articles about the waiting list for public housing,
          ***Sydney Morning Herald* 1990–99**

quality, high housing costs or outright homelessness. They contained implicit messages of support for public housing. Most of the articles written about the waiting list for public housing in the latter period were much more simple and factual, recording the length of the waiting list but providing little context for it. This was consistent with the discursive shift in reporting that will be discussed in more detail below.

The next theme of articles concerned general policy debates about public housing. By far the largest collection of articles in this category related to the political commitment of governments in providing public housing (see Figure 7.4). The distribution of these articles is best explained by conflicts between the state and federal governments over the level of funding for public housing. Claims by the federal Labor government that the state Liberal government was insufficiently committed to public housing inspired most of the articles in the early period. The articles in 1995 (the only year when the party in power in NSW at state level was the same as at federal level) were in response to the Labor governments at both levels attacking their Liberal oppositions over levels of commitment to public housing. In addition welfare groups attacked proposed changes to housing policy outlined by the federal Labor government (in its *Community and Nation* statement). The peak of articles about political commitment came in the period after the election of the federal Liberal government in 1996, going through into 1997. The state Labor government mounted sustained attacks upon proposed federal changes to public housing policy. This was clearly evident in the press releases from the office of NSW Housing Minister Craig Knowles. Debates about political commitment are interesting because they highlighted the central tension

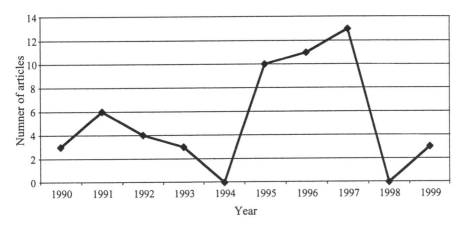

Figure 7.4   Total articles concerning the political commitment of government to public housing, *Sydney Morning Herald* **1990–99**

apparent in the discursive shift elaborated in this paper. Articles attacking the political commitment of either level of government in relation to public housing contained implicit and explicit messages about the significance of public housing as a welfare measure. The problem as constructed by those attacking the political commitment of government was that public housing was undervalued, and undersupplied as a result. From 1995 onwards however the retort, from the federal government in particular, was that public housing policy had failed, and that a more flexible alternative could be provided by offering rent assistance for the private market. In 1998 there were no articles about the political commitment of government to public housing. This was because the state Labor government shifted its political attack to one concerned with the impacts of the Goods and Service Tax (GST) on public housing provision (a trend noted by Putnis, 2001 in his more general work on welfare reporting), a strategy clear in ministerial press releases. Again the implicit message from state government in this period was that public housing was a necessary welfare measure, which would be undermined by federal government policy changes.

The final theme of articles concerned those where there was an explicitly positive coverage of issues in public housing. Interestingly the biggest sub-theme of articles that reported on public housing in a positive fashion were historic discussions of public housing. Unlike the articles mentioned earlier which focus on the failures of past management, these articles expressed nostalgia for a time when public housing places were more 'ordinary' places where families lived.

These articles then were tinged with criticism of contemporary policies, and the implicitly dysfunctional neighbourhoods they have created.

One of the most interesting characteristics of these articles however is the dominance of one journalist, in producing positive articles. The public housing articles analysed here were produced by a large number of journalists from different sections of the newspaper. The journalist who produced the most articles about public housing was Adele Horin. While Horin was responsible for only 6 per cent of the total articles analysed, she wrote over 25 per cent of all the explicitly positive articles about public housing, identified here under the sub-theme of 'Cost effective and stable'. Other articles written by Horin, but not included in this theme strongly carried the implicit message that public housing was a necessary and undersupplied welfare measure. The sources chosen by Horin were part of the framing of her articles, and more strongly favoured sources from the housing, community and welfare sectors than sources used by other journalists in the collection. In producing opinion pieces for the *Sydney Morning Herald*, her articles, more clearly than most, were positioned in a way that signalled her own ideological positioning. Known to be a journalist supportive of government intervention in welfare (Putnis, 2001), Horin railed against the position on public housing advanced by other journalist members of her interpretive community, as the more detailed discussion of a discursive shift below reveals.

Discursive Shift: Key Moments

In this section I draw on the key elements of the argument presented above, that a discursive shift in the representation of public housing in NSW took place in the *Sydney Morning Herald* in the 1990s. This discursive shift was contested and incomplete, but nonetheless public housing was increasingly reported as a policy failure, and a new policy solution, providing rent assistance for the private market was advocated as an appropriate policy response. Here I highlight the nature of these contests by drawing on some key articles from three periods of the 1990s.

1990

At the beginning of the decade the predominant message of newspaper articles was that public housing was a necessary welfare measure that was undersupplied. The relatively high number of articles about the public housing waiting list during this period implicitly projected this message. The message was carried through in headlines such as 'No Place To Go' (Lumby, 1990), and the framing of articles with key statements such as 'the pool of public housing is drying up' (Lumby, 1990). The understanding of public housing as necessary and undersupplied framed the interpretation of events for journalists. Thus Dixon (1990) wrote: 'At a time when families are made to wait as long as five years for public housing, the Housing Commission of NSW has put a block of flats on the market for sale as a development site'. Events relating to public housing were interpreted through a problem frame emphasising that a key problem with public housing was that there was an insufficient supply and that the government should act to solve the problem.

There was however another strand to the discourses about public housing during this period; public housing places, characterised in the media as large estates, were dysfunctional. Such a representation was particularly apparent in 1990 following murders in public housing in Surry Hills. An article by Deborah Cornwell exemplified such reporting. The headline of the article 'Dumping Ground for Misfits' brought together the two key features of this discourse. First, that public housing allocation policy was inadequate; it created 'dumping grounds'. Second, people who lived in public housing were perceived as 'other'; they did not fit with mainstream society. Cornwell (1990) (and many other journalists and politicians talking about the issue) framed an understanding of the murder in the following place-based terms:

> What happened in Surry Hills today is just what you would expect to happen when you put people with mental and social problems into the same place with no social support. They just don't cope and you can't expect them to (Cornwell, 1990).

While such a framing was particularly stark following the dramatically newsworthy events of the murders, it was also a clear part of much of the wider reporting on the theme of the social characteristics of residents and their neighbourhoods. Such an understanding of public housing places as dysfunctional plays a vital role in the recasting of public housing as a failed policy, as is clear from an analysis of debates about public housing in the middle of the 1990s.

1995–96

Three clear features emerge from an analysis of discourses around public housing in the period of 1995–96. The first is that there is an even stronger emphasis on the depiction of public housing places as 'failed'. The language of these years is stronger than in 1990, with estates being described in a number of articles as 'ghettos'. Second, this understanding of public housing places helps journalists situate proposed changes to public housing policy as a solution to failed policies of the past. Third, the strident rebuttal of this point of view by some journalists (notably Adele Horin) reflects the extent to which this new understanding of public housing was challenging an entrenched notion of public housing as a necessary welfare measure.

Public housing places were rendered as dysfunctional ghettos in a number of different contexts during 1995 and 1996. The following quotations provided a flavour of such renditions:

> As far as Bob and his ilk are concerned, Claymore is an urban ghetto ruled by the law of the fist (Allison and Jopson, 1995).

> The NSW government is already turning to the private sector for new sources of public housing, giving tenants an alternative to the 'ghetto' estates built in the 1960s and 1970s (Millet, 1995b).

Proposed policy changes that would shift federal government funds towards rent assistance were presented in the light of these representations of public housing as failed. But there was another dimension to the failure of public housing; it provided too much assistance to those tenants able to gain access to public housing. Like council tenants in the UK (Jacobs, Manzi and Kemeny, 2003b), then, public housing tenants came to be represented as overly privileged. The following quotations reveal this new framing of the problems of public housing in a way that advocated rent assistance as the 'common sense' solution to the problem.

> Fundamentally the system fails where it appears to succeed. Public housing in Australia provides decent accommodation for the frail, vulnerable and poor in society, but this only creates a disparity between unlucky low-income people who have missed out on one of the 370,000 public housing places, catering for about 900,000 people (Cleary, 1996).

> A radical plan by the Federal Government to overhaul the $2.5 billion public housing system will increasingly by-pass failed estates in favour of private leased dwellings with greater access to employment opportunities (Millet, 1995a).

Such views of the policy change did not go uncontested. The NSW Housing Minister appeared in the press throughout 1996 commenting on different aspects of how the proposed changes in policy would impact upon public housing tenants, and emphasising the importance of public housing as a welfare measure. The most strident rebuttals of the notion of public housing as failed came in the articles of Adele Horin. During 1995, at the key moment when representations of public housing were being most contested Horin wrote a strong and direct defence of public housing under the headline 'Public Housing is a Foundation of Society'. Cresswell (1996) argues that the most direct statements result from incidents that transgress the normalised assumptions of individuals. Horin's passionate defence of public housing indicated that the nature of the debate, and the position of some of her journalist colleagues were transgressing her assumptions about public housing. The following quotes demonstrate these points:

> home ownership has kept old people from the poorhouse and children from vagrancy. Public housing extended similar benefits to low income people – the secure, affordable base on which to build a life (Horin, 1995).

> Public housing gets bad press – and much is wrong with some old estates in bad locations. But it's the best chance very-low income families will ever get to provide their kids with a secure start in life (Horin, 1995).

In these quotations Horin acknowledged that there were some problems with public housing. However, she rejected the notion that the solution to these problems was to abandon public housing; rather it was to improve public housing.

The middle of the 1990s was a period when the nature of a 'public housing' problem was being contested in the pages of the *Sydney Morning Herald*. Sources and journalists within the interpretive community that had input into producing these articles were clearly divided about what they saw as the key elements of the public housing problem. Increasingly however, articles came to be framed around the notion that if there were problems with public housing, there was a new solution to address these problems.

1999

Adele Horin continued to highlight the position that public housing was a necessary welfare measure in her articles until the end of the decade. Drawing on the dominant implicit position of most newspaper articles from the beginning of the decade she drew attention to the problems of people on the waiting list for public housing thus:

There is a public housing waiting list of 96,000. And over the past six years, there has been a massive increase in the numbers of homeless who need a bed (Horin, 1999b).

More public housing is needed; research has shown public housing gives governments more bang for their buck (Horin, 1999a).

The headlines of Horin's articles, 'Calls For Help ... But No-one Receiving' and 'A Great Australian Dream in Danger of Extinction', framed Horin's view that her understanding of public housing was being marginalised, both in the pages of the *Herald* and in public debate in general. The state government continued to position public housing as an essential service, but one that was being undermined by changes to federal funding. As the decade ended then there was still debate about the 'problem' of public housing, and solutions to this problem. However, the coverage of public housing as a news item was increasingly sporadic, and many more articles about 'problems' in public housing were framed around notions of rent assistance as the solution to the problems. This narrative framing was heavily influenced by the role of the federal government in constructing their housing solution as one that addressed both the 'problem' of public housing estates and the privilege of public housing tenants.

Conclusions

At the beginning of the millennium debates around public housing in Australia were at the crossroads. In NSW, as this chapter has revealed, there was continuing debate about whether public housing was a necessary welfare measure, or the creator of failed housing estates that simultaneously created a class of privileged tenants. Future research is needed to explore the extent to which the debates revealed here were peculiar to the NSW context, or more broadly indicative of debates around public housing occurring in other parts of Australia.

Jacobs, Kemeny and Manzi (2003a) note that there are many competing narratives circulating around housing at any given time. They draw on Hajer's (1993) test of what narratives will prove enduring. Hajer (1993) argues that narratives that endure will come to dominate discursive space and will be reflected in institutional practices. In applying this test to the discourses examined here, the outcome for public housing remains uncertain. While the 1990s certainly saw the rise of a discourse championing the solution of rent assistance, discourses of public housing as a necessary welfare measure remained. The practice of policy reflected this position. For as Caulfield (2000) and Dodson (2001) note, while public housing suffered from reductions in funding in real terms, and rent assistance for the private market increased dramatically in this period, public housing remains an important source of affordable housing for people on low incomes in Australia. Despite the small size of the public housing sector there is still considerable support for the maintenance of public housing among welfare groups and Labor state governments. The debates of the 1990s will no doubt continue in the lead up to the renegotiation of the Commonwealth-State Housing Agreement in 2007. There is little doubt that a switch to rent assistance for the private market is consistent with the neoliberal agenda of the Federal Liberal Government of Prime Minister John Howard (Caulfield, 2000). Whether such a switch can be more thoroughly embedded in institutional change in future years remains to be seen.

This chapter has also demonstrated the utility of analysing media coverage of housing debates. Understanding media articles as socially constructed by interpretive communities allows a window into how policy debates become translated into the public forum, and how political actors attempt to present their version of policy problems as common sense. An understanding of newspaper articles as produced by interpretive communities also allows us to move beyond simply looking at these articles as producing and reproducing dominant discourses to, examine whether and how different groups are able to access the media in order to have their opinions heard. Moreover as the analysis presented here shows, the work of particular journalists in an interpretive community may be vital in presenting alternative reactions to and interpretations of policy change. Such an analysis requires a consideration of newspaper coverage of housing issues across a long time period, so that the articles produced in the heat of a policy debate can be contextualised within a broader understanding of a policy.

Acknowledgements

The author would like to thank Phil O'Neill and Robyn Dowling who both, though in quite different ways, provided inspiration for this chapter. The author would also like to thank Nick Nolan for excellent research assistance on the paper. The author also acknowledges the helpful comments of Pauline

McGuirk, Keith Jacobs, Brett Hutchins, Elaine Stratford, Julie Davidson, Chris Gibson, Natalie Moore, Sally Lane and Kris Ruming that contributed to the development of the ideas in the paper. The author would also like to thank the staff of the NSW Parliamentary Library for their assistance in accessing ministerial press releases.

Notes

1   The statistics in this paragraph are drawn from Darcy and Randolph (1999), NSW Department of Housing (1999a, 1999b, 2000, 2001).
2   Federally the Australian Labor Party (ALP) governed Australia from 1990 to March 1996. The Liberal-National Party Coalition (the two major conservative parties of Australian politics) governed for the rest of the decade. In New South Wales the Liberal-National Coalition governed from 1990 to March of 1995, the ALP from March 1995 to the end of the decade.
3   Provided by the State Library of New South Wales.
4   Altheide (2000) describes a similar technique for developing a comprehensive understanding of media articles as qualitative media analysis.

References

Allison, C. and Jopson, D. (1995), 'Suburb Living in Fear', *Sydney Morning Herald*, 21 November.
Altheide, D. (1996), *Qualitative Media Analysis*, London: Sage.
Altheide, D. (1997), 'The News Media, the Problem Frame, and the Production of fear', *The Sociological Quarterly*, Vol. 38, No. 4: 647–68.
Altheide, D. (2000), 'Tracking Discourse and Qualitative Document Analysis', *Poetics*, Vol. 27, No. 4: 287–99.
Atkinson, R. (2000), 'Narratives of Policy: The Construction of Urban Problems and Urban Policy in the Official Discourse of British Government 1968–1998', *Critical Social Policy*, Vol. 20, No. 2: 211–32.
Berkowitz, D. and TerKeurst, J. (1999), 'Community as Interpretive Community: Rethinking the Journalist-source Relationship', *Journal of Communication*, Vol. 49, No. 3: 125–36.
Brindle, D. (1999), 'Media Coverage of Social Policy: A Journalist's Perspective', in B. Franklin (ed.), *Social Policy, the Media and Misrepresentation*, London: Routledge: 39–50.
Burgess, J. (1985), 'News from Nowhere: The Press, the Riots and the Myth of the Inner City', in J. Burgess and J. Gold (eds), *Geography, the Media and Popular Culture*, Sydney: Croom Helm: 192–228.
Caulfield, J. (2000), 'Public Housing and Intergovernmental Reform in the 1990s', *Australian Journal of Political Science*, Vol. 35, No. 1: 99–110.
Cleary, P. (1996), 'Fifty-one Years and Still Trying to get the Housing in Order', *Sydney Morning Herald*, 23 September.
Cornwell, D. (1990), 'Dumping Ground for Misfits', *Sydney Morning Herald*, 31 August.

Cresswell, T. (1996), *In Place/Out of Place: Geography, Ideology and Transgression*, Minneapolis: University of Minnesota Press.

Darcy, M. and Randolph, B. (1999), *Strategic Directions for Housing Assistance: Final Report Prepared for the NSW Department of Housing*, Sydney: Urban Frontiers Program, University of Western Sydney.

Deacon, D. (1999), 'Charitable Images: The Construction of the Voluntary Sector', in B. Franklin (ed.), *Social Policy, the Media and Misrepresentation*, London: Routledge: 51–68.

Dixon, A. (1990), 'North Sydney Site may Sell for up to $1.5m', *Sydney Morning Herald*, 19 June.

Dodson, J. (2001), *The Order of Housing Things: Public Housing Policy Discourse in New Zealand and Australia, 1983–1999*, unpublished thesis submitted in total fulfilment of the degree of Doctor of Philosophy, Faculty of Architecture, Building and Planning, University of Melbourne.

Driedger, S. and Eyles, J. (2003), 'Charting Uncertainty in Science Policy Discourses: The Construction of the Chlorinated Drinking Water Issue and Cancer', *Environment and Planning C: Government and Policy*, Vol. 21, No. 3: 429–44.

Duncan, S., Edwards, R. and Song, M. (1999), 'Social Threat or Social Problem? Media Representations of Lone Mothers and Policy Implications', in B. Franklin (ed.), *Social Policy, the Media and Misrepresentation*, London: Routledge: 238–52.

Entman, R. (1991), 'Framing US Coverage on International News: Contrasts in the Narratives of the KAL and Iran Air Incidents', *Journal of Communication*, Vol. 41, No. 4: 6–27.

Entman, R. (1993), 'Framing: Towards a Clarification of a Fractured Paradigm', *Journal of Communication*, Vol. 43, No. 4: 51–8.

Entwhistle, V. and Sheldon, T. (1999), 'The Picture of Health? Media Coverage of the Health Service', in B. Franklin (ed.), *Social Policy, the Media and Misrepresentation*, London: Routledge: 118–34.

Fairclough, N. (1995), *Media Discourse*, London: Edward Arnold.

Forte, J. (2002), 'Not in my Social World: A Cultural Analysis of Media Representations, Contested Spaces and Sympathy for the Homeless', *Journal of Sociology and Social Welfare*, Vol. XXIX, No. 4: 131–84.

Fowler, R. (1991), *Language in the News: Discourse and Ideology in the Press*, London and New York: Routledge.

Franklin, B. (1999), 'Introduction', in B. Franklin (ed.), *Social Policy, the Media and Misrepresentation*, London: Routledge: 1–14.

Gaber, I. (2000), 'Government by Spin: An Analysis of the Process', *Media, Culture and Society*, Vol. 22, No. 4: 507–18.

Goodchild, B. and Cole, I. (2001), 'Social Balance and Mixed Neighbourhoods in Britain since 1979: A Review of Discourse and Practice in Social Housing', *Environment and Planning D: Society and Space*, Vol. 19, No. 1: 103–21.

Hajer, M. (1993), 'Discourse Coalitions and the Institutionalisation of Practice: The Case of Acid Rain in Britain', in F. Fischer and J. Forester (eds), *The Argumentative Turn in Policy Analysis and Planning*, London: University College London Press: 43–76.

Hastings, A. (1998), 'Connecting Linguistic Approaches and Social Practices: A Discursive Approach to Social Policy Analysis', *Journal of Social Policy*, Vol. 27, No. 2: 191–211.

Horin, A. (1995), 'Public Housing is a Foundation of Society', *Sydney Morning Herald*, 15 December.

Horin, A. (1996), 'Housing Plan may add to Homelessness', *Sydney Morning Herald*, 14 September.

Horin, A. (1999a), 'A Great Australian Dream in Danger of Extinction', *Sydney Morning Herald*, 4 December.

Horin, A. (1999b), 'Calls for Help ... But No-one Receiving', *Sydney Morning Herald*, 17 March.

Jacobs, K. (2001), 'Historical Perspectives and Methodologies: Their Relevance for Housing Studies', *Housing, Theory and Society*, Vol. 18, Nos 3–4: 127–35.

Jacobs, K. and Manzi, T. (1996), 'Discourse and Policy Change: The Significance of Language for Housing Research', *Housing Studies*, Vol. 11, No. 4: 543–60.

Jacobs, K., Kemeny, J., and Manzi, T. (2003a), 'Power, Discursive Space and Institutional Practices in the Construction of Housing Problems', *Housing Studies*, Vol. 18, No. 4: 429–46.

Jacobs, K., Kemeny, J., and Manzi, T. (2003b), 'Privileged or Exploited Council Tenants? The Discursive Change in Conservative Housing Policy from 1972–1980', *Policy and Politics*, Vol. 31, No. 3: 307–20.

Jones, M. and Ward, K. (2002), 'Excavating the Logic of British Urban Policy: Neoliberalism as the "Crisis of Crisis Management"', *Antipode*, Vol. 34, No. 3: 473–94.

Lindlof, T. (2002), 'Interpretive Community: An Approach to Media and Religion', *Journal of Media and Religion*, Vol. 1, No. 1: 61–74.

Lumby, C. (1990), 'No Place to Go', *Sydney Morning Herald*, 11 January.

Marston, G. (2000), 'Morality, Myth and Metaphor: A Critical Discourse Analysis of Public Housing Policy in Queensland', *Critical Social Policy*, Vol. 20, No. 3: 349–73.

Marston, G. (2002), 'Critical Discourse Analysis and Policy Oriented Housing Research', *Housing, Theory and Society*, Vol. 19, No. 2: 82–91.

Mee, K.J. (2000), *Creating the West: A Dialogical Analysis of Representations of Western Sydney in the 1990s*, unpublished PhD thesis, University of Sydney.

Miewald, C. (2002), 'National Myths, State Policy, and Community-directed Media: Representational Politics and the Reconfiguration of Welfare', *Urban Geography*, Vol. 22, No. 5: 424–39.

Miller, D. (1993), 'Official Sources and "Primary Definition": The Case of Northern Ireland', *Media, Culture and Society*, Vol. 15, No. 3: 385–403.

Millet, M. (1995a), 'Revamp for Public Housing', *Sydney Morning Herald*, 23 September.

Millet, M. (1995b), 'States Told to Fix Housing Crisis', *Sydney Morning Herald*, 28 November.

Myers, G., Klak, T. and Koehl, T. (1996), 'The Inscription of Difference: News Coverage of the Conflicts in Rwanda and Bosnia', *Political Geography*, Vol. 15, No. 1: 21–46.

NSW Department of Housing (1999a), *Directions for Housing Assistance: Beyond 2000 Background Paper*, Sydney: NSW Department of Housing.

NSW Department of Housing (1999b), *NSW Department Corporate Plan*, Sydney: NSW Department of Housing.

NSW Department of Housing (2000), *Community Renewal: Transforming Estates into Communities, Partnership and Participation*, Sydney: NSW Department of Housing.

NSW Department of Housing (2001), *NSW Department of Housing Annual Report 2000–2001: Working Together*, Sydney: NSW Department of Housing.

Oatley, N. (2000), 'New Labour's Approach to Age-old Problems: Renewing and Revitalising Poor Neighbourhoods – the National Strategy for Neighbourhood Renewal', *Local Economy*, Vol. 15, No. 2: 86–97.

Papadakis, E. and Grant, R. (2001), 'Media Responsiveness to "Old" and "New" Politics Issues in Australia', *Australian Journal of Political Science*, Vol. 36, No. 2: 293–308.

Parisi, P. and Holcomb, B. (1994), 'Symbolizing Place: Journalistic Narratives of the City', *Urban Geography*, Vol. 15, No. 4: 376–94.

Pfau, M. (1995), 'Covering Urban Unrest: The Headline Says it All', *Journal of Urban Affairs*, Vol. 17, No. 2: 131–41.

Putnis, P. (2001), 'Popular Discourse and Images of Poverty and Welfare', in R. Fincher and P. Saunders (eds), *Creating Unequal Futures? Rethinking Poverty, Inequality and Disadvantage*, Crows Nest: Allen and Unwin: 70–101.

Sabatier, P. (1987), 'Knowledge, Policy-oriented Learning, and Policy Change', *Knowledge: Creation, Diffusion, Utilization*, Vol. 8, No. 4: 649–92.

Sabatier, P. (1993), 'Policy Change over a Decade or More', in P. Sabatier and H. Jenkins-Smith (eds), *Policy Change and Learning: An Advocacy Coalition Approach*, Boulder, CO: Westview Press: 13–39.

Schudson, M. (1996), 'The Sociology of News Production Revisited', in J. Curran and M. Gurevitch (eds), *Mass Media and Society*, London: Arnold: 141–59.

Stone, D. (1989), 'Causal Stories and the Formation of Policy Agendas', *Political Science Quarterly*, Vol. 104, No. 2: 281–300.

Van Dijk, T. (1988), *News as Discourse*, Hove and London: Lawrence Erlbaum Associates.

Williams, K. (1999), 'Dying of Ignorance? Journalists, News Sources and the Media Reporting of HIV/AIDS', in B. Franklin (ed.), *Social Policy, the Media and Misrepresentation*, London: Routledge: 69–86.

Zelizer, B. (1993a), 'Has Communication Explained Journalism', *Journal of Communication*, Vol. 43, No. 4: 80–88.

Zelizer, B. (1993b), 'Journalists as Interpretive Communities', *Critical Studies in Mass Communication*, Vol. 10, No. 3: 219–37.

Chapter 8

Organisational Research: Conflict and Power within UK and Australian Social Housing Organisations

Michael Darcy and Tony Manzi

Introduction

An important, albeit neglected, aspect of social constructionism is its grounding in interpersonal, institutional and organisational interaction (Kemeny, 2002: 140). This interactionist basis to housing theory is often obscured in the strong emphasis on discourse analysis found in much of the constructionist literature and can lead to considerable misunderstanding (if not distortion) of its objectives. Ironically an obsession with discourse as text has served to obscure many important insights of discourse theory. Discourse therefore needs to be seen as interactive, indeed dialectical – text in context. Thus, the dominance of an uncritical approach which only focuses on the meaning of specific discourses, may not allow a strong focus on wider structural factors. For example, it enables writers such as Somerville and Bengtsson (2002) to criticise constructionism for failing to advance beyond the urban managerialist debates of the 1970s and for failing to consider the wider structural constraints upon housing management staff (131). A useful distinction can therefore be made between discourse analysis in general and 'critical discourse analysis' (CDA). Thus van Dijk (2001) contends that CDA focuses on social *problems* and *issues*. Rather than *describe* discursive practices, CDA tries to *explain* them in terms of social interaction and structure; additionally it focuses on the ways in which discourse structures enact, confirm, legitimate, reproduce, or challenge relations of power and dominance in society (see also Fairclough and Wodak, 1997).

Constructionism in fact is integrally bound up with consideration of institutional and organisational constraints within a context of interpersonal interaction (Kemeny, 2002: 140). This emphasis places questions of language as central to the definition of interests, attitudes, values and norms. However, crucially it does not ignore other structural features which examine how relations of power, interest groups and inequality are exercised. Critical discourse analysis is premised on an understanding of discourse as not just language, or even text, but text in its political and cultural context, and especially on the influence of each upon the other.

Consequently, the current chapter attempts to consider some intersections with the agency-structure debate, by using housing management practices in two different contexts as a case study. It draws upon qualitative research methods and ethnographic data from studies in Australia and England. It utilises insights from constructionist theory and discourse analysis to demonstrate some of the possibilities of an interpretive social science which can analyse the working of conflict and power within the social setting of contemporary housing organisations. Although this chapter does not claim to be a substantial interactionist study, it aims to highlight a general approach that organisational researchers can take. Before considering the case study material, the chapter first outlines some methodological questions.

Methodology in Housing Research

Quantitative studies, which seek to establish positive knowledge of housing problems, based on empirical evidence, and to provide justification for the development of rational solutions have been the bread and butter of housing research for decades. However, in the context of the post-structuralist and latterly postmodernist turns in social science, such methods have often been denigrated for their appeal to grand narratives of modernism and the tendency to define social needs and problems in terms of the available solutions. The 'linguistic' and 'cultural' turns in social theory criticise the privileging of the broad structural view of academics and bureaucrats and the diminishing of the housing experience of ordinary homeowners, tenants or homeless people. For example, even survey methods put the researcher at an advantage in framing questions and in some cases predetermining available answers.

On the other hand, the open ended, qualitative methods preferred by post-structuralist social scientists are claimed to empower research subjects by allowing them directly to define; that is to attach meaning and value, to their own housing experience. Nonetheless, such approaches, however well theorised they may be, are often dismissed for their particularity and their ineffectuality to policy makers. That is to say, their findings are often conditional and not readily able to be generalised.

Most importantly, it is also true that individual experience cannot be isolated from social context and that the meaning and value that individuals assign to experience is conditioned by, and interpreted through, the existing social order (at the local, national and global levels). The grammatical and lexical constraints of the social order, which we can call 'societal orders of discourse' (Fairclough, 1992) shape what is expressed by research subjects, how it is expressed, and how it is understood and interpreted by researchers and policy makers. As we shall show, three significant trends in the overarching orders of discourse, namely 'technology', 'commodity' and 'democracy', can be seen to have influenced housing management in Britain and Australia. These categories are based upon (and further develop) Fairclough's analysis; they represent processes and directions of change in discourse at the highest orders or changes in the 'conditions of possibility' of discourse. Crucially, they tend to limit what can be said or the terms in which things can be spoken of.

In this context, housing research and housing policy can be viewed as text-based social practices. Research and policy texts are the medium through which housing problems and solutions are constructed, as they are in turn constructed within societal orders of discourse. This process of meaning-making through texts and associated social practices (text production, reading, interpreting, resisting or implementing) is a rich and valid site for housing research, which provides a bridge between structuralist and post-structuralist paradigms of housing research, but only when such texts and practices are critically analysed within their social and organisational context.

The underlying tenet of research which draws on notions of social and/ or discursive construction of social phenomena is that actors (including researchers) do not merely provide descriptions of events, but that they, and the descriptions they produce, are constrained within and are themselves constitutive of, wider policy discourses and conflicts. This understanding demands from researchers themselves a much greater emphasis on reflexivity, and an acceptance that structural and material processes are a necessary but not sufficient explanation for contemporary housing problems. In organisational terms this model of research sees a field of study as a cooperative intellectual practice, with a tradition of historically produced norms, rules, conventions and standards of excellence that remain subject to critical debate, and with a narrative content that situates and gives meaning to it (Reed, 1993: 176). The narrative of contemporary housing management is a story of bureaucratic failure, state withdrawal and an increasingly authoritarian approach. At the same time, there is a strong focus on consumer sovereignty and democratic participation.

Within an organisational context the importance of rhetoric and persuasion is hugely significant given the highly political context of housing management. It is therefore curious that despite the interest in linguistic and cultural models of housing management, there has been such a paucity of constructionist influenced empirical management research.

Social Constructionism and Housing Organisations: An Interactionist Perspective

The roots of an interactionist trend in housing theory can be found in symbolic interactionism (Blumer, 1969) emphasising the importance of negotiated order in the resolution of conflict and compromise (Strauss, 1978). As Kemeny argues, within housing studies there has been an almost total absence of genuinely interactionist studies in the constructionist tradition (2002: 141). The work of Fine (1984) or Hallett (2003) on organisational cultures and a substantial literature on the conflicts and negotiations involved in generating organisational settings has not permeated discussion of organisational change within a housing context. Drawing on the influence of Strauss (1978) such studies claim that what takes place within institutions is a complex process of bargaining between different interest groups (for example Silverman, 1975; Silverman and Jones, 1976; Scanzoni, 1979 ; O'Toole and O'Toole, 1981). The neglect of this literature within housing is surprising given the widely accepted view of housing management as an essentially contested practice, subject to substantial indeterminacy of roles and definitions (Franklin and Clapham, 1997). The negotiated order model of housing organisations would therefore understand organisational behaviour as the outcome of interactions between social groups and their mutually developed meanings. Constructionism can therefore help to bridge micro and macro distinctions aided by the application of symbolic interactionism (Collins, 1981). Sometimes criticised for its neglect of structural features, in contrast, an interactionist dimension to constructionism helps to overcome some of the weaknesses of traditional and relativistic models of social construction.

The social constructionist tradition involves a questioning of 'common-sense' or 'taken-for-granted' explanations of reality. In particular constructionist methodologies imply a reaction against 'foundational' assumptions, which contain the view that certain kinds of experience are ingrained or essential. For social constructionism the active process of observation takes place within the realm of language. Hence 'whatever does exist we can only know by way of our constituting it through discourse' (Grint, 1995: 8). Discourse and language are therefore of central importance in understanding how we perceive and make sense of the social world, indeed organisations are essentially discursive constructions, so that our understanding entails a process of active and reflexive engagement.

Discourse analysis is therefore strongly relevant to studies of organisational change, which has been an increasingly important topic for housing management research in recent years. Housing managers can best understand and 'manage' change (and resistance) by application of a paradigm which accounts for the subjective and constructed nature of meaning in organisations.

An understanding of, and a critical approach to, discourse in housing policy and housing management can play a key part in the mediation of contradictions which bedevil housing organisations, whilst a failure to adopt the reflexivity which such an approach demands is often a central factor in policy failure. The following examples, from housing management in New South Wales, Australia and England, demonstrate that the articulation of management objectives and development of strategies to achieve them is far from a rational exercise based on positive knowledge; yet the dominance of a rational modernist organisational paradigm is seen to prevent key players from understanding and responding to complex situations.

To understand the process of contemporary housing management, it is necessary to explore the ways in which particular kinds of knowledge are valorised. For example, contemporary organisations are likely to pay particular attention to issues of accounting, creativity and entrepreneurialism. Whilst being careful not to make the mistake of drawing conclusions on the basis of limited studies within just two countries, what can be seen are that following Fairclough (1992), within a housing context, three specific tendencies in the orders of discourse of contemporary capitalist society have assumed particular prominence in the period since the late 1980s. These are the discourses of *technologisation, commodification* and *democratisation*.

The first central discourse of *technologisation* or 'professionalisation' presents management as a generic and rational function. In the terms of modern public management, there is no distinction between and public and private organisations. Management is viewed as a neutral activity, removed from the vagaries of political conflict, which was the *sine qua non* of municipal, or public, housing management. There is a surprising degree of convergence in policy goals and across policy fields. Examples of these trends include an emphasis on particular skills; for example accounting and performance management. Managerialism is closely linked to the project of measuring efficiency and neoliberal models of the political economy of welfare. Housing organisations are expected to demonstrate a high degree of expertise in management skills of innovation and creativity in pursuit of measurable efficiencies, understood as the antithesis of traditional bureaucratic organisations. Traditionally housing managers have had limited success in establishing themselves as a professional grouping (Furbey et al., 2001: 37). Consequently, contemporary practitioners were keen to stress that their skills and abilities were to be placed upon a more sound foundation of organisational practices. Thus Chief Executives of UK housing associations welcomed the post-1988 environment as part of a 'brave new world' of housing finance, often presenting a view of the manager as a heroic individual, welcoming potential threats as opportunities. The influence of private finance and reduced public subsidy led to the emergence of a highly competitive culture where elitism amongst the largest developing organisations was encouraged.

However, one of the paradoxes of the technical discourse is that innovation coexists with a high degree of regulation and centralised control. This leads to one of the myths of management control seen in the introduction and abolition of compulsory competitive tendering (CCT) in Britain in the 1980s. This policy was designed to increase organisational efficiency (and by implication to hasten privatisation) through imposing competitive practices upon municipal bureaucracies. By forcing organisations into a separation between client and contractor sections and for the latter to tender for their work it was assumed that other (private) service providers would win the contracts. However, in practice most of the contracts were awarded to 'in-house' teams. Hence a policy that was presented as a solution to bureaucracy in practice was found to result in increased levels of bureaucratic monitoring and control (Walsh, 1995).

The implementation of the Housing Act 1988 in Britain was a significant milestone in the transformation of the social rented sector. The Act placed local authorities in an 'enabling role'; moving from providers to facilitators of housing provision. The emergence of housing associations as the main housing providers in place of local authorities was designed to end the municipal monopoly on social rented housing and to allow a more pluralistic housing sector to materialise. However, in practice, the Act has seen the emergence of a small number of elite organisations constituting a new form of professional monopoly controlling increasing proportions of the housing association stock (Housing Corporation, 2002). This elitist tendency is a further indicator of the professionalisation of expertise, a phenomenon which is also evident in New South Wales. The bias towards larger organisations went hand in hand with an increased centralisation of policy, as the Housing Corporation was increasingly able to exercise its authority over policy decisions (Malpass, 1999: 31). Thus, a policy ostensibly designed to improve flexibility and pluralism had the opposite effect. Consequently, members of staff commented on the way that hierarchical trends were re-emerging within the sector.

A further example of the technologisation of housing management discourse from another context can be found in Australia around the issue of estate regeneration and social mix. In 2002 and 2003 the New South Wales (NSW) Department of Housing was embroiled in a dispute with tenants and community groups over plans to redevelop a large broad-acre estate in Western Sydney. In the 1960s and 1970s, greenfield development of public housing on the fringes of major cities was seen as the solution to problems of housing affordability and poor housing quality, especially in inner cities. Thirty years later, large-scale public housing estates were seen, by policy makers and housing managers as problems rather than solutions. Increased targeting of allocations and consequent residualisation and stigmatisation of the tenant population, combined with poor locational amenity have resulted in the management practices of housing authorities being blamed for the 'social exclusion' of public housing tenants.

Reams of research reports were commissioned, and many conferences convened, in an attempt to address this new 'problem': however, the dominant approach has been to attempt to measure the problem, to discover positive knowledge about it and to develop general solutions based on this knowledge – that is to say, a 'rationalist' approach. But, in general terms, few questioned the definition of the problem, or bothered to ask for whom it is a problem. Much effort was put into pilot programmes and evaluations, where approaches included: diversification of management through stock transfers; neighbourhood improvement schemes such as fencing and upgrading; community development projects including tenant councils and community gardens; place management and intensive tenancy management schemes in defined areas; and various combinations of these.

In the 1,400-unit Minto estate, 40 kilometres southwest of the centre of Sydney, most of these approaches were mounted on a small scale at different times over the past decade. While many were seen as very successful and positive experiences by tenants and local housing managers, and a few, including an intensive tenancy management project, have been held out as examples of good practice and given awards by central Housing Department officials, none were extended or continued largely due to cost. In this and other estates, for example, neighbourhood improvement demonstration projects have successfully impacted on vandalism, antisocial behaviour and turnover. But in each of these estates, the 'improved' areas, stood in contrast to surrounding unimproved streets.

Nonetheless, over the years of being paid such policy attention, Minto has seen a marked reduction in street crime, a drop in rehousing applications, and a dramatic increase in resident participation in a variety of community activities – including a significant project to repurchase and redevelop local open space as a community park. However, in its latest response to the emerging social exclusion discourse in social research and social policy, the Department of Housing apparently concluded that the 'real' solution to the 'problem' lies in increased 'social mix', which in turn can only be achieved through 'tenure mix'. Because of the existing stigma surrounding Minto in particular and public housing in general, owner occupiers are not expected to buy dwellings in the area (at a price which will finance its revitalisation) unless the estate is progressively demolished and totally redeveloped in a joint venture with the private sector. This solution is based on a particular, and highly political, reading of limited and often speculative research. Yet the power of management discourse to rationalise a particular narrative linking 'problem' and 'solution' has seen it proceed quickly to implementation.

With the bulldozers threatening, tenants mounted a campaign to promote their own perspective, which reflects the fact that they did not attach the same meanings to aspects of life in Minto as did the social exclusion researchers. Indeed, especially in recent times and through various community development activities, residents felt very connected to the local community and, ongoing amenity issues notwithstanding, cited their common experiences as a positive feature of estate life. Many residents expressed a fear of 'social mix' as a threat to community solidarity. Moreover, most were faced with the threat of relocation to another area at some time during the 10-year redevelopment program, and many refused to leave. In the example housing management can be seen as a technologised discourse which excluded those it was ostensibly designed to assist. Interestingly, the tenant groups resisting the redevelopment of this estate have chosen a positivist, quantitative research design to present their perspective in political debate. They have explicitly recognised the orders of discourse impinging on their ability to redefine their lives and the situation in which they find themselves.

The second trend in societal orders of discourse is that of *commodification*. The notion of choice, which is closely associated with commodification through market discourse, was again presented as one of the solutions to the problem of bureaucracy and planning. Commodification involved a move to introduce market-based mechanisms (Walsh, 1995) within public sector institutions. It also attempted to apply the insights of public choice theory, in particular to counter the effect of 'bureau-maximisation' (Dunleavy, 1991), namely that the prime incentive for public managers is to win support for higher levels of public spending. A further component of commodification in the guise of 'new public management' (NPM) was the influence of new institutionalist economics, emphasising the importance of principles such as contestability, user choice, transparency and incentive structures (Williamson, 1985). The final strand of a commodification discourse in NPM was the notion of entering an era of 'post-bureaucratic' organisations (Hoggett, 1991) whereby traditional ideas of hierarchical management structures were seen as outdated. The 1988 reforms have led to the development of a risk culture, with much greater use of private finance and a more pluralistic approach to service provision. This was reflected in interviews with senior managers who welcomed the opportunities that the new organisational culture would bring.

The commodification discourse primarily conceptualises welfare recipients as consumers, equivalent to retail customers. With public organisations such as local authorities constructed as problems in themselves, due to their monopoly status, the problem of social housing is conceptualised as a loss of autonomy. The discourse of choice therefore offers the opportunity of freedom, independence and transparency. Within a British context, examples of this choice discourse include the 1980 Right to Buy (RTB) for council tenants and are witnessed clearly in the Housing Act 1988 which marked the culmination of the Thatcherite reform programme with voluntary sector housing associations chosen as the preferred vehicles for the realisation of government objectives. The Housing Act 1988 'reflected Government's determination to transform rented housing from a poor quality local authority near monopoly to a diversified, privately funded and managed business sector' (Glennerster et al., 1991: 398). Superficially the rationale was to ensure a more flexible and locally based response to housing policy in line with libertarian aspirations. However, the realisation of the programme was to result in quite different outcomes. For local authorities, the objective was to play a strategic, enabling role, removing political control from the management of housing policy and allowing the voluntary sector to be the main provider of new social housing. At the same time, the late 1980s also saw the flowering of a programme of modernisation under the rubric of a 'new public management' (NPM) which was partly based on individualistic philosophies and involved a strong focus on the concepts of 'disaggregation', 'incentivisation' and 'competition' (Dunleavy and Hood, 1994). As an example the initiative of 'choice-based' lettings (taken from the so-called *'Delft'* model based on housing allocation systems in Holland) provides a more recent example of this discourse. The *Delft* model provides opportunities for residents to express choices about their preferences for accommodation; for landlords to advertise openings in local press and for more information to be provided about available options. As well as increased transparency, in theory, this initiative provides much greater choice in the process of letting accommodation. However, the reality of available opportunities within local authority areas of high demand mean that in practice the choices are highly constrained and may lead to increased frustration amongst resident groups.

Similarly, since the mid-1980s a programme of stock transfer from local authorities to alternative landlords has gained momentum. A lack of funding opportunities available to municipal housing providers has made the transfer programme increasingly attractive to local authority decision-makers. Although couched in a language of choice and with proposals dependent upon a ballot of residents, in practice there is normally a clear 'preferred option' that local authority landlords strongly encourage residents to vote for. As will be shown in an Australian context, such initiatives can lead to a reduction rather than an increase in choice.

As a manifestation of this commodification, in a familiar scenario, capital funding for development of public housing in New South Wales (NSW) has, in real terms, been in decline for almost two decades. This is an element of wider retrenchment of state welfare and has been accompanied by the emergence of a discourse, linked to neoliberalism, which strongly rejects large scale intervention by bureaucratic government agencies, in development and consumption of, amongst other things, housing. (Public housing in Australia, while never making up more than 6 per cent of all housing, has been dominated for more than half a century by large state-run providers.) In response, small non-government housing associations began to assert their potential as alternative housing managers. Originally, promoters of 'community housing' sought the use of such organisations as a means of expanding overall investment in social housing by moving some of the social housing debt burden outside the government sector, however this has not happened due to government reluctance to provide tax incentives to private investors.

Despite this, 'community housing' has grown, not through new investment, but at the expense of traditional 'public housing'. Some government funds have been provided directly, and the management of existing public housing stock has been transferred to housing associations. In NSW the Department of Housing stated that 'the aim of expanding the community housing sector is to increase consumer choice to be able to offer alternatives for people whose only housing options in the past may have been public housing, and to promote innovation in service delivery through increased competition in the sector' (NSW Dept of Housing, 1999). The language of 'consumer choice', innovation and competition is borrowed from neoliberal discourse and appeals to an individualised form of agency which clearly breaks with the modernist notion of a singular rationality and generalised solutions to social problems. In its implementation, however, this programme demonstrates the persistence of old organisational paradigms and the dangers of resisting a more reflexive approach.

In the majority of cases, housing units to be transferred were first 'decanted' (the Department of Housing's own term) and upgraded; however, in the drive to achieve policy targets, a number of tenanted properties were also earmarked for transfer by Departmental officers. In a study conducted at the height of the first wave of transfer activity in 1999 and 2000, Darcy and Stringfellow (2001) interviewed tenants and housing managers in five areas where transfers of tenanted properties had been completed affecting about 100 tenants.

This research documents many examples of where tenants' choices were severely restricted, or their rights threatened, in the name of increasing the size of the community sector and as a way of improving tenants' choices. In summary, sitting tenants were given a very limited choice in that they were advised that the management of their house was to be transferred, and that they could either transfer their tenancy or be relocated. In some cases tenants were told that if they failed to agree to the transfer, they would have no right to be rehoused. Most of the tenants interviewed had very little information and believed that the whole process was taking place because the Department could not afford to maintain their houses, which was ironic given that all houses were upgraded *before transfer* as an incentive to the housing associations to accept them. In four of the five case study areas all tenants accepted the transfer.

In their analysis the researchers cited this as an example of the bureaucratic phenomenon described by Weber as 'goal displacement'. Once the unitary policy objective had been operationalised through procedures and targets, the management machinery was set in train to achieve them regardless of the fact that the overarching discursive rationale for the policy, increased tenant choice, was being severely compromised, and oblivious to the evident need for a more individualised strategy – or at least incapable of mounting such a strategy through its existing organisational form. The study criticised both the Department and housing associations for basing decisions and actions on property considerations rather than the needs or wishes of their tenants (or 'consumers').

Most telling however, was not the finding of the research itself, but the reaction of senior departmental officers to its planned publication. After being sent a courtesy draft, the Department produced a lengthy critique of the research methodology and threatened to publicly challenge its conclusions on the basis of lack of rigour. Their central criticism concerned sampling method and sample size. While the research had involved interviews with only 17 tenants (20 per cent of affected tenants in the five areas identified for study) and 19 housing managers, no statistical techniques were attempted in relation to their responses and no claims were made that their experiences could be generalised. The study had explicitly set out to record individual and area case studies, and the report was dominated by verbatim reporting of interviewee responses. The report provided ample evidence that individual tenants had experienced the whole process as a loss of choice, and that this had happened fairly systematically across examples in five different housing estates. Indeed in the one area in the study where tenants *had* been involved before management transfer arrangements were finalised, they had been able to exercise choice and all opted against the transfer. The Department's inability to respond constructively to knowledge (in this case, about its own operations) generated through qualitative methods, and presented from the viewpoint of tenants, as opposed to positivist responses to predetermined policy 'problems', thus becomes a site for productive research and critical discourse analysis. In this case the original 'problem' constructed as a lack of 'consumer choice' and 'innovation' arose not from tenants' experience, but from the dominant neoliberal critique of government. Having subsequently been reinterpreted through bureaucratic management practices into a set of performance targets, the 'problem' could be solved by an essentially discursive device of 'transferring management' to a non-government organisation.

The final order of discourse identified is that of *democratisation*. Within a housing context this is presented as a discourse of *community* or *neighbourhood*, and also manifests in more formal requirements for consultation, participation, and more transparent policy debate. One of the most significant contemporary trends in British housing management has been the emphasis on resident involvement, seen through initiatives such as Best Value (DETR, 1999a) and Tenant Participation Compact (DETR, 1999b) schemes which impose a statutory duty on landlords to consult with resident groups. Despite these initiatives, interviews with residents continue to demonstrate strong views that the rules of the game were set by professionals who continue to use strategies to exclude unwelcome opinions or conflicting views.

The social problem was often conceptualised by managers as a problem of homogeneity and social exclusion. The central policy problems that government has to tackle are those of antisocial behaviour and lack of responsibility. Solutions therefore about revolve around increasing social capital; developing sustainable communities through the adoption of multiple tenure strategies; resident participation and partnership. Contemporary examples include the implementation of 'neighbourhood management' and policies to tackle antisocial behaviour (sometimes presented as a problem of moral behaviour). Landlords are therefore encouraged to develop policies of social engineering. However, there is also a more paternalistic strand to this discourse, marking a return to Poor Law distinctions between deserving and undeserving groups. As examples policies to implement 'probationary' or 'introductory' tenancies in Britain or 'renewable' tenancies in Australia show the commonality of these disciplinary welfare models. In Britain, proposals to limit housing benefit payments to tenants in breach of their tenancy agreements also illustrate this increasingly authoritarian streak in housing management as well as conflicting with the discourse of democratisation. Residents were therefore expected to take responsibility for own behaviour heralding the end of what some have termed a 'dependency culture'. The former Social Security Minister in Britain, Frank Field (2003) has published a polemic against the behaviour of social housing tenants, calling for police officers and housing managers to become surrogate parents in the absence of traditional parental controls.

At the same time that these hierarchical trends were emerging, creating a new type of organisational bureaucracy, others were stressing the importance of organisational roots, in particular a strong sense of social justice, that housing associations needed to retain their roots as campaigning organisations. Managers within housing associations were often keen to retain their sense of identity as value-based, locally situated, community organisations. In one sense the endeavour to place housing management in the hands of non-government organisations, where tenants supposedly have more influence, is an attempt to present social housing as a more democratically organised institution. However, this shift is clearly at odds with the more powerful trends of technologisation and commodification, which find more resonance (or interdiscursivity) within the neoliberal discourses on reform and modernisation of the state.

The examples illustrate the dynamic contradictions within housing policy seen from British and Australian perspectives. The tensions between technologisation (seen in rationalist and professionalised management);

commodification (exercised in consumer power through competition and choice) and democratisation (through consultation, participation and tenants' rights) result in a range of unforeseen consequences. Housing management practice is thus subject to continuous interpretation and redefinition reflecting the efforts and abilities of various players to reconstruct the goals and objectives of policy in their own interests. The power of the various interest groups will prove conclusive in their ability to gain a foothold on the policy agenda. The examples considered, illustrate how the negotiated orders of housing management showing how organisational dominance and conflict are played out within the context of housing policy.

The trajectory of housing policy in Britain and Australia shares a number of key discursive features. Influenced by a neoliberal ideology, the reform programmes were initially aimed at ending the bureaucratic dominance of the post-war period, whereby local or state authorities were the planners, managers and developers of social housing. The above trends represent three different directions that housing organisations have encountered before and after 1988. These tendencies in societal orders of discourse indicate the way in which power and conflict relationships are played out within a housing management setting. The rules of the game continue to be set largely by professionals and organisational elites, despite paying lip service to concepts of democratisation and resident involvement.

Conclusion

One of the main strengths of a constructionist approach to management theory is to illustrate the complexity of organisational behaviour. By making explicit the underlying ideological themes and the extent to which these are contested and contradictory we can reach a much more useful understanding of the way in which conflicts within organisations appear and may be resolved. The alternative is to see resistance to organisational change as part of a pathological unwillingness to embrace change for the benefit of vested interests (an assumption which underpinned much of the reform programmes of the 1980s).

What these examples illustrate is the importance of a more subtle understanding of the social composition of organisational culture, reliant on qualitative analysis in order to understand the way that organisational change is routinely interpreted in a variety of different ways. Studies of organisational behaviour need to beware of the perils of modernist remedies, assuming that managerialism can function as a panacea for the wrongs of bureaucracy. The lesson of these experiences for housing managers and policy makers is not just that they should diversify the type of research they commission to inform their decisions, although that is clearly true. More important is that they need a research approach which allows, and indeed encourages, analysis of the way in which policy problems are constructed, interpreted and received. Such an approach will encompass non-positivist, interpretive epistemology and methods, but will also be based upon a critical and reflexive sociological framework.

As Hood (2000) has shown there is a tendency within management thought for ideas to appear and reappear constantly, reinventing themselves as new remedies to hitherto insoluble problems. These trends can be seen to have continued with governing administrations placing their faith in a regeneration policy reliant upon supposedly 'pragmatic' or managerial solutions under the mantra 'what matters is what works' (Social Exclusion Unit, 2001).

Managerialist discourse in housing relies heavily on redefining 'tenants' as 'consumers' or 'customers' who are rational and well informed (Marston, 2000). This, commodified construction of social housing relationships in turn implies a need for democratisation of discourse, opening policy and decision making to a wider range of interpretations. On the other hand, housing managers are presented as possessing the technologies for determining and delivering consumer preferences in the most efficient way. Given the historically powerless culture of public housing tenancy, and the fact that, for the most part, government continues to be the paying customer of housing organisations, housing organisations and housing managers can expect to lead a contradictory existence.

In particular by combining quantitative data about housing policy with detailed qualitative research, a more useful picture of the different tensions within housing policy can be identified. The use of critical discourse analysis and social interactionist theory also allows an identification of the way in which definitions of the problem of social housing are dependent on the power of interest groups (Jacobs et al., 2003). In addition by examining case studies from different cultural backgrounds a research agenda can be developed to understand the extent to which global trends may be evident within approaches to housing policy. Whilst acknowledging the importance of common themes such as technologisation, democratisation and commodification, it is important not to overstate the similarities between two distinct organisational environments with different institutional backgrounds. The negotiated order of housing organisations therefore illustrates some crucial similarities as well as contrasts.

A decision to label something as a social problem is inevitably ideologically loaded and dependent on a specific narrative of events. Once accepted it can become incorporated into an institutional custom and thereafter forms an accepted part of professional practice. An interactionist approach shows how future research can interrogate different elements of public management; for example by analysing how problems in social housing are defined and by questioning the methods which governments and practitioners use to justify their interventions.

References

Blumer, H. (1969), *Symbolic Interactionism: Perspective and Method*, Englewood Cliffs, NJ: Prentice Hall.

Clapham, D. (1997), 'The Social Construction of Housing Management Research', *Urban Studies*, Vol. 34, Nos 5–6: 761–74.

Collins, R. (1981), 'On the Microfoundations of Macrosociology', *American Journal of Sociology*, Vol. 86, No. 5: 984–1014.

Darcy, M. and Stringfellow, J. (2001), *Tenants' Choice or Hobson's Choice: A Study of the Transfer of tenanted dwellings from Public Housing to Community Housing in NSW*, Sydney: Urban Frontiers Program, University of Western Sydney.

Department of the Environment, Transport and the Regions (DETR) (1999a), *Best Value in Housing: a Guide to Tenants and Residents*, Housing Green Paper London: DETR.

Department of the Environment, Transport and the Regions (DETR) (1999b), *National Framework for Tenant Participation Compacts*, London: DETR.

Dunleavy, P. (1991), *Democracy, Bureaucracy and Public Choice*, London: Harvester Wheatsheaf.

Dunleavy, P. and Hood, C. (1994), 'From Old Public Administration to New Public Management', *Public Money and Management*, Vol. 14, No. 3, July–September: 9–16.

Fairclough, N. (1992), *Discourse and Social Change*, London: Polity Press.

Fairclough, N. and Wodak, R. (1997), *Critical Discourse Analysis*, in T. van. Dijk (ed.), *Discourse as Social Interaction*, London: Sage: 258–84.

Field, F. (2003), *Neighbours from Hell*, London: Politico's.

Fine, G. (1984), 'Negotiated Orders and Organisational Cultures', *Annual Review of Sociology*, Vol. 10: 239–62.

Franklin, B. and Clapham, D. (1997), 'The Social Construction of Housing Management', *Housing Studies*, Vol. 12, No. 1: 7–26.

Furbey, R., Reid, B. and Cole, I. (2001), 'Housing Professionalism in the United Kingdom: The Final Curtain or a New Age?', *Housing, Theory and Society*, Vol. 18, Nos 1–2: 36–49.

Glennerster, H., Power, A. and Travers, T. (1991), 'A New Era for Social Policy: A New Enlightenment or a New Leviathan?', *Journal of Social Policy*, Vol. 20, No. 3: 389–414.

Grint, K. (1995), *Management: a Sociological Introduction* Cambridge: Polity Press.

Hallett, T. (2003), 'Symbolic Power and Organizational Culture', *Sociological Theory* (June), Vol. 21, No. 2: 128–49.

Hoggett, P. (1991), 'A New Management in the Public Sector?', *Policy and Politics*, Vol. 19, No. 4: 243–56.

Hood, C. (2000), *The Art of the State: Culture, Rhetoric and Public Management*, Oxford: Clarendon Press.

Housing Corporation (2002), *Housing Associations in 2001: Performance Indicators*, London: The Housing Corporation.

Jacobs, K., Kemeny, J. and Manzi, T. (2003), 'Power, Discursive Space and Institutional Practices in the Construction of Housing Problems', *Housing Studies*, Vol. 18, No. 4: 429–46.

Kemeny, J. (1992), *Housing and Social Theory*, London: Routledge.

Kemeny, J. (2002), 'Reinventing the Wheel? The Interactional Basis of Constructionism', *Housing, Theory and Society*, Vol. 19, Nos 3–4: 140–41.

Malpass, P. (1999), 'No Strings Attached?', *Roof*, July/August: 30–31.

Marston, G. (2000), 'Metaphor, Morality and Myth: A Critical Discourse Analysis of Public Housing Policy in Queensland', *Critical Social Policy*, Vol. 20, No. 3: 349–73.

New South Wales (NSW) Department of Housing (1999), *Directions for Housing Assistance Beyond 2000*, Background Paper, NSW Department of Housing.

O'Toole, T. and O'Toole, A. (1981), 'Negotiating Interorganisational Orders', *The Sociological Quarterly* (winter), Vol. 22, No. 1: 29–41.

Reed, M. (1993), 'Organisations and Modernity: Continuity and Discontinuity in Organisation Theory', in J. Hassard and M. Parker (eds), *Postmodernism and Organisations*, London: Sage.

Scanzoni, J. (1979), 'The Centrality of Negotiation to the Study of Social Organisation', *Contemporary Sociology*, Vol. 8: 528–30.

Silverman, D. (1975), 'Accounts of Organisations: Organisational Structure and the Accounting Process', in J. McKinlay (ed.), *Processing People: Cases in Organisational Behaviour*, New York: Holt, Rinehart and Winston: 269–302.

Silverman, D. and Jones, J. (1976), *Organisational Work: The Language of Grading/the Grading of Language*, London: Collier Macmillan.

Social Exclusion Unit (2001), *Preventing Social Exclusion*, London: The Stationery Office.

Somerville, P. and Bengtsson, B. (2002), 'Constructionism, Realism and Housing Theory', *Housing, Theory and Society*, Vol. 19, Nos 3–4: 121–36.

Strauss, A. (1978), *Negotiations: Varieties, Contexts, Processes and Social Order*, San Francisco: Jossey-Bass.

Van Dijk, T. (2001), 'Critical Discourse Analysis', in D. Schriffin, D. Tannen and H. Hamilton (eds), *The Handbook of Discourse Analysis*, Oxford: Blackwell.

Walsh, K. (1995), *Public Services and Market Mechanisms: Competition, Contracting and the New Public Management*, London: Macmillan.

Williamson, O. (1985), *The Economic Institutions of Capitalism*, New York: Free Press.

Chapter 9

Social Constructionism and International Comparative Housing Research

Anna Haworth, Tony Manzi and Jim Kemeny

Introduction

The purpose of this final chapter is to consider how a social constructionist approach can be applied to the analysis of international comparative housing research and to point out some key issues to consider in the analysis of cross-cultural questions. Much of the comment is therefore necessarily speculative. We argue that many difficulties in conducting international comparative housing research might be diminished by the use of a social constructionist approach and that important lessons for such an approach can be found in the methods used by ethnologists and social anthropologists.

There are a number of advantages of comparative research, and especially of international comparative research. Firstly, it allows cross national evaluation to be made, avoiding simplistic judgments based upon national assumptions. Secondly, it enables an explanation of social change placed within specific contexts. Thirdly, it explores the impact of global processes, particularly within Western European states as a consequence of policy harmonisation and convergence. More generally it can broaden horizons of research, learning lessons from other cultural contexts. It is particularly important within the context of increasing attempts in the European Union to apply common solutions to what are seen as common problems through the process of policy transfer.

However, what is generally absent from many comparative studies in housing – whether international or not – is both an explicit acknowledgement of the limitations of comparative study, particularly on methodological grounds and the theoretical tools to overcome these limitations. Many of these limitations relate to terminological differences, interpretative complexities and different sedimentations of knowledge. One of the main problems in international comparative research is the unintentional imposing of values and motives on foreign cultures, and interpreting their policies and social organisation from unexplicated and taken-for-granted standpoints. This partly arises because of the essentially positivist approach taken in many of the studies. It is also based upon common assumptions – mostly unquestioned – about experiences, knowledge and the accumulated history of housing processes.

Even though housing researchers have in recent years increasingly used explicit theoretical assumptions as the basis of their work (Kemeny and Lowe, 1998) the inadequacies of many comparative studies in housing (already pointed out by Harloe, 1985: xiv) continue to be apparent. The reasons for that inadequacy are discussed in Harloe's study and include the difficulty of arranging long periods of study in other countries, the lack of availability of information, and differences in methods of data collection. So although comparative housing researchers are aware of the problem of ethnocentric bias, the response tends not to go further than carrying out research on other countries by nationals of that country, or at least by the use of 'ethnic' informers to correct the ethnocentric interpretations of the researcher.

Despite a growing influence in housing research (Clapham, 1997; Franklin and Clapham, 1997; Jacobs and Manzi, 2000; Somerville and Bengtsson, 2002) constructionism has not yet made an impact in comparative policy research. It can do so in two main ways.

First and foremost, a central aim of social constructionism is to avoid deterministic views of social structure as an external and fixed entity that cannot be changed by purposeful and coordinated action. On the contrary, social constructionists argue that social structure, including policy, are the products of conscious human agency and therefore highly malleable. Social constructionists therefore focus on how power struggles between vested interests can have a decisive impact on what policies are implemented and that outcomes are by no means foregone conclusions. Much can be learned from comparative studies of housing policies that refuse to adopt a Whig version of history and instead show how different outcomes of power struggles are not only possible but have produced significant international differences in housing policy and provision.

Second, social constructionism can help to avoid ethnocentrism. One of the strengths of the social constructionist perspective is the conscious use of reflexivity to question ethnocentric taken-for-granted assumptions and to treat foreign societies superficially similar to our own as 'anthropologically strange' (Garfinkel, 1984). In striving to achieve this, the use of the ethnographic method has been of central importance. The real difficulty comes when trying to apply ethnographic methods to large-scale processes of social change and policy formation.

The Value of the Social Constructionist Approach to Comparative Research

As mentioned in other chapters, at the heart of a social constructionist approach is the questioning of assumptions, in particular, the questioning of 'objective' definitions of social reality, and indeed, the categorisation of different aspects of the social world as social problems (Jacobs et al., 2003). Central to these arguments is the notion that such definitions or categorisations are dependent on the ability of different interest groups to impose their own agenda. The objective of such research is to explore where power lies in terms of dominant interests. Where societies have different structures, or different power formations, there is likely to be variance in the location of power. The resultant dominance of different 'problems' between one society and another is the result of that distribution. The centrality of language in this approach is critical when different languages are being used and therefore there is a need to explore meanings from the perspective of different cultural viewpoints, taking into accounts values, assumptions and traditions, using native speakers wherever feasible.

Despite an emphasis on qualitative research, interview data and 'taking actors seriously' (Somerville and Bengtsson, 2002) there is still a neglect of the interactive dimension in comparative study. By situating questions of language centrally within the definition of interests, attitudes, values and norms, social constructionism emphasises both the discourse and wider structural features which explain how relations of power, interest groups and inequality are exercised.

Moreover, international comparative housing research has not generally been based on an intimate knowledge or a phenomenological feel for the societies being compared and has not generally integrated the ideological and theoretical implications of the study areas. Instead much international comparative research continues to rely on highly descriptive accounts of different countries; an abstracted empiricism that has failed to learn from the sociological, political science and public policy literature. In addition, international comparative housing research has not paid sufficient attention to historical factors that can enable detailed contextual studies to be undertaken.

As shown in previous chapters (for example see Travers, Chapter 2 and Kemeny, Chapter 4) social constructionism emphasises the importance of negotiated order in the resolution of conflict and compromise (Strauss, 1978). This emphasis on power relationships and integration of structure and agency has not been sufficiently explicit in much of the comparative housing literature. An examination of how power is exercised over time and in different cultural contexts, can help to explain continuity and change in social customs and traditions and how this affects actors differentially (Granovetter et al., 1993; Pierson, 1994). This approach can assist in the explanation of differences and similarities between countries as well as helping us to understand why particular institutions sustain themselves – or indeed why they change – over time.

In particular, an anthropological approach to housing can deepen knowledge of societies and situate the place of housing, geography and locality within cultural contexts. For example, Granovetter's discussion of the concept of 'social embeddedness' (1985) illustrates the way in which economic activities are deeply implicated in wider social structures. Embeddedness refers to the contingent nature of economic action with reference to 'cognition, culture, social structure and political institutions' (Zukin and Di Maggio, 1990: 15).

Granovetter's work on the social construction of the electricity industry (Granovetter et al., 1993; Granovetter and McGuire, 2004) is a prime example of how treating as anthropologically strange existing and often taken-for-granted social forms can yield important social constructionist insights into the formation of large-scale organisations. Granovetter asks why the electricity industry developed into a highly centralised form of production requiring its own distribution network. Why generate electricity centrally in huge power stations and then distribute it by wire on pylons and poles over hundreds of miles to hundreds of thousands of individual buildings? Why didn't electricity production take the form of the manufacture of white goods electricity generators to be installed in each home, much like refrigeration, or, indeed, much like early forms of electricity production using windmills and watermills? There are many ways to organise electricity production. The answer lies in the entrepreneurial innovation of developers and the backing they received from financiers, politicians, scientists, builders and other interests.

In similar fashion Latour (1986) demonstrates how the concept of scientific objectivity, technical expertise and neutrality is constructed in a laboratory environment. Relying on power relationships between doctors, nurses and statisticians, scientific facts are constructed through the social interaction between these groups as much as discovered through experimentation. Much the same goes for housing. For example, dualist rental systems in the US are part of an ongoing struggle between vested interests. The house-building industry has played the dominant role in determining housing policy through mobilising an effective coalition of financial institutions, construction and real estate interests. In conjunction with neoliberal federal governments hostile to public expenditure *per se*, public housing was successfully redefined as a highly residual and stigmatised tenure, whilst subsidies were given to developers through generous housing loans (Heidenheimer et al., 1975: 78).

Social constructionism allows the researcher to question taken-for-granted realities by refusing to see existing social organisation as inevitable and understanding them in terms of how they were constructed out of interaction, negotiation and conflict. Rather than viewing economic behaviour in timeless and asocial utilitarian terms as a consequence of rational individuals abstractly exercising choice, it is seen as integrally bound into constraints based on interpersonal relations.

Furthermore, in order to understand the way in which comparisons can be made in housing and social policy, its methodology needs to be steeped in the 'lifeworld'. In particular the crucial question to consider is how do people construct their lifeworld? This requires *in situ*, contextualised studies of locale through the methods of social anthropology.

The particular problems of cross-national research are acknowledged to be a fundamental issue in social anthropology, 'transnational studies pose unique methodological problems' (Russell Bernard, 1998: 23) – and, it might be argued that in social anthropology there is a greater awareness of the dangers of ethnocentrism and particularly of the danger of imposing incorrect interpretations on different cultures as a result of a failure to understand the meanings that actors bring to social settings in different countries. An ethnographic approach such as that espoused by anthropology and social anthropology might lead to 'thicker' descriptions of the outcomes of housing policies and practice in different countries.

The interpretive methods advocated in anthropology are rare in comparative housing although the examination of housing systems in other countries has many similarities with an anthropological investigation. Writing about ethnographic description, Geertz notes that, 'what we call our data are really are own constructions of other people's constructions of what they and their compatriots are up to' (Geertz, 1993: 9). Further, he argues that anthropological writings are what he terms second and third order interpretations. By definition, only a 'native' can make first order interpretations: 'it's his culture' (Geertz, 1993: 15). Notwithstanding that second order interpretations may be useful in understanding activities in other countries, the notion that such interpretations are perhaps less reliable or valuable than first order interpretations should be acknowledged in comparative housing studies.

Where such problems are acknowledged these have generally been in terms of definitional problems rather than genuinely interactionist studies. As

Schutz (1962) notes 'the typifying medium *par excellence* by which socially derived knowledge is transmitted is the vocabulary and syntax of everyday language' (Schutz, 1962: 14). This notion links to Blumer's idea of a knowable empirical reality (Blumer, 1969). In anthropology, the notion of learning the 'field language' is emphasised in research methods texts (for example Russell Bernard, 1998: 146).

The idea from cultural anthropology that 'key informants' will give an entrée to the society being studied (Russell Bernard, 1998) is illustrated in William Whyte's study, where 'Doc' was his key informant, introducing him to the group, interpreting their behaviour, explaining group behaviour and so on (Whyte, 1993). The use of 'informants' although not expressed as such, is mentioned in Harloe's study of comparative housing systems where he notes that:

> to understand the development of social housing in the six countries with which we have been concerned, it was as important to identify and trace the significance of some general political and economic changes in all advanced capitalist countries as it was to grasp the nationally specific circumstances in which these changes were experienced and which shaped the responses to them (Harloe, 1995: 528).

In order to help him achieve the latter, he acknowledges a debt of gratitude to, 'five leading students of housing in their own countries' (Harloe, 1995: 528), as having responded to his requests for information and having acted as consultants during the course of the research. That this contact was 'managed' by the researcher stands in contrast to the traditional approach followed in anthropology which is, perhaps, closer to that of participant observation. In a study of the latter, Jorgensen notes that, 'developing and sustaining relationships with insiders in the field is crucial to gathering accurate and dependable information. This process is not unlike being socialised into a way of life' (Jorgensen, 1989: 80).

How can these insights be applied to a social constructionist method within housing study? We suggest that the application of social constructionist theory and ethnomethodolical approaches can illustrate the dangers in an unreflective approach to contemporary housing initiatives. The problems of second order interpretations can be most usefully addressed through a focus on three primary areas.

Translation, Transferability and Transfer in Housing Policy

Following Tizot (2001) we identify three specific aspects to cross-national study that can be usefully addressed within a constructionist framework. These relate to the *translation* of discourse, *the transferability* or adaptability of ideas and concepts and policy *transfer* or implementation.

First, issues around the *translation* of housing discourse have been inadequately addressed in much housing literature. The accumulated history of certain concepts makes translation problematic (Eyraud, 2001). That inadequacies may originate from ethnocentrism has been acknowledged by some authors undertaking comparative housing studies. For example, Hallet (1988) in his study of land and housing policies in Europe and the US argued that, 'a person writing about his own country tends implicitly to assume an understanding of institutions and concepts which may be completely alien to outsiders. The consequent confusions are frequently compounded by problems of language' (Hallet, 1988: 1).

The notion of social exclusion provides a good recent example of an 'essentially contested concept' (Levitas, 1996). As Kleinman maintains:

> the concept of social exclusion has moved to centre stage with remarkable speed. ... in doing so, its lack of precise meaning has been an asset rather than a liability ... but in trying to understand problems, and to devise effective policy, this lack of precision is more than a liability – it is catastrophic (Kleinman, 2000: 53).

The popularity of social exclusion as an explanatory concept lies in its ability to refer to a number of different and competing notions; for example the rhetoric of social justice; of community development and of an underclass. These different discourses help to justify very different policy responses: from community development to an interventionist and authoritarian housing management response. Drawing a boundary between acceptable and unacceptable behaviour the social exclusion discourse distinguishes between those inside and outside and can be applied in very different ways in different social contexts.

The concept emanated in a French context in the 1970s and is widely assumed to have had general application across Western European societies. However, the use of the phrase often neglects the essential differences in the treatment of the concept. Social exclusion in Britain is used as a notion to describe a range of factors:

a short hand term for what can happen when people or areas suffer from a combination of linked problems such as unemployment, poor skills, low income, poor housing, high crime, bad health and family breakdown (Social Exclusion Unit, 1997).

The significance of this concept is that it enables governments to adopt a managerial sense of purpose; providing 'joined-up solutions to joined-up problems'. Principally driven by a concern for social order (Byrne, 1999) the social exclusion discourse can be used to illustrate the way in which concepts of 'dynamism, relativity and agency' are used to justify different policy interventions by Western governments under the guise of strategies to tackle social exclusion.

From a French perspective, there is much greater emphasis on the social aspects of social exclusion. Writing about 'la malaise des banlieues', Stébé noted the spiral of exclusion and precariousness in the suburbs, and the implication that these were economic at root, but argued that social relations were often part of the explanation, citing family breakdown, and poor relations between different social groups. Discussing the process of social marginalisation (*disqualification sociale*) he sees this as beginning with fragility of employment, proceeding to dependence on social welfare and the final phase, 'rupture', encompassing distance from the sphere of work, breakdown of family relations, absence of housing, no stable income, health problems and so on (Stébé, 1999: 51), concluding that, 'elle est le résultat d'une série d'échecs aboutissant á une importante marginalisation [it is the result of a series of handicaps leading to significant marginalisation]' (Stébé, 1999: 51). Examining the accumulation of 'handicaps', he addresses the issue spatially noting that the areas *défavorisés* are far from city centres, comprise poorly constructed, high rise buildings, where almost a third of inhabitants are less than 20 years old, twice as many households have six or more persons than in the wider population, one third fewer people are over 65, three times as many are immigrants than elsewhere, and unemployment is almost twice the national average: a homogeneous underclass, *mal-vivre*, characterising the areas as 'territoires de misère' (Stébé, 1999: 54).

In contrast to the area approach favoured by British governments, Stébé argues against the notion of suburbs as ghettos, stating that the inhabitants of the *Habitations á Loyer Modéré* (HLM) areas are no more than the first victims of a major sociological shift. This shift is provoked by the decline of the working class, by the effects of unemployment on individuals, and by social and economic cleavage, benefiting those in stable employment, with cultural capital, and disadvantaging those who find themselves relegated to *la galère* (the margins), without qualifications, excluded from work and deprived of housing (Stébé, 1999: 54). These sociological definitions have a strong resonance in a country with a republican tradition; where concepts of solidarity and social cohesion play a central role in welfare policy (Silver, 1995; Burchadt et al., 2002: 2).

Somewhat closer to the UK approach is that favoured by the EU, whose policy document on social exclusion states that, 'employment is the best safeguard against social exclusion', advocating policies to develop employability and social protection schemes, decent and sanitary housing, basic services, and recommending a focus on exclusion from the information society (Council of the European Union, 2002). Other societies choose to translate social exclusion differently, for example in Scandinavia the term is used to refer to 'the poorest of the poor' (Abrahamsom and Hansen, 1996, cited in Burchadt et al., 2002: 3).

Other terms can be translated in many different ways. Thus, a study comparing housing systems in the US and Britain noted that 'direct comparisons of the adequacy of housing supply between the two countries are fraught with both conceptual and data comparability problems' (Karn and Wolman, 1992: 16). This study noted the differences in the way in which 'households' were defined in Britain and America, for instance. Additionally, Harloe discussed the comparability of the term 'private rented' and found that:

> the German housing market fits least comfortably into the broad classification adopted here ... the social housing sector is not confined to rental housing or to a particular category of builders or landlords (Harloe, 1985: 63).

Definitions of social housing in Britain convey subsidised housing, managed by public landlords, whilst in Germany distinctions between public and private are much more obscure, 'not confined to rental housing or a particular category of builders or landlords' (Harloe, 1985: 63).

An example of a misinterpretation of a concept in relation to social housing is the use of the notion of 'The People's Home' or *folkhemmet* in Sweden (Harloe, 1995). This has carried a strong emblematic status but in its housing dimension:

> has retained an ambiguous and shifting status on the margins of the welfare state, the least decommodified and the most market-determined of the conventionally accepted constituent elements of such states (Harloe, 1995: 2).

However, the concept of *Folkhemmet* includes the whole welfare state and not just housing; it refers to the whole of society in which everyone helps one another without distinction. It uses the metaphor of the milltown where the capitalist *patron* is implicitly replaced by the egalitarian socialist collective. In a famous Per Albin Hannson speech of 1928 he says: 'In the people's home there are no step-children'. He hijacks the conservative mill-town ideal but replaces the arbitrary discretionary welfare dispensed by the patron with an Owenite socialist community. This demonstrates the problems inherent in translation, where a generalist description of social structure is used as if indicative of housing policy in isolation.

Other areas of housing policy have proved resistant to translation, for example in many countries there is no equivalent to the housing management profession found in the UK. These examples illustrate the way in which the problems of translation, which are at root theoretical difficulties; based on discourse, relate to more practical issues, for example about professionalism and the practice of housing organisations.

The second issue, that of *transferability* relates to the adaptability and comparison of cross-national policy. This question raises important issues about where and when policies can be most usefully applied within different contexts. These issues are highly relevant within debates about policy harmonisation and integration. There is an increasing emphasis on these cross-national initiatives at a European level, particularly as increasing levels of research activity and funding are commissioned by the European Union. Yet there is little explicit debate about these questions. What is often neglected is the specificity of member states, for example how the notion of a 'social market' differs between different European societies. Similarly, how social democratic countries will tend to have more overtly collectivist institutions than social market economies (Hutton, 2002: 264). The assumption appears to be that there are inherent similarities, resting upon convergence models as policy harmonisation becomes increasingly important. However, the contextualisation and history of different member states is often ignored in the eagerness for research to be commissioned and comparability to be demonstrated.

The use of the term 'social capital' provides one example of these difficulties in transferability. This notion derives from the work of Putnam (2001) but is used in very different ways in the US and Britain, as well as in different European countries. Whilst a review of the claims being made for social capital in contemporary writing notes that, 'common to all these definitions of social capital are the concepts of reciprocity and generalised trust'

(Johnston and Percy-Smith, 2003: 324); it has also been noted that, 'overuse and imprecision have rendered it a concept prone to vague interpretation and indiscriminate application' (Forrest and Kearns, 2001: 2137). In the US it can be taken to justify a neoliberal model of state withdrawal, whereas in Britain it is more commonly used to expressed a 'third way' notion that combines, public, private and voluntary activity.

It is widely acknowledged that there is no common definition of the idea of social capital. For the World Bank, for example, social capital:

> refers to the internal social and cultural coherence of society, the norms and values that govern interactions among people and the institutions in which they are embedded. Social capital is the glue that holds society together and without which there can be no economic growth or human well-being (World Bank, 2001).

A report on 12 projects funded by the Social Capital Initiative, launched in 1996, notes, 'strong evidence that social capital is a pervasive ingredient and determinant of progress in many types of development projects, and an important tool for poverty reduction' (World Bank, 2001: xi), claiming that social capital, 'can have a major impact on the income and welfare of the poor by improving the outcome of activities that affect them' (World Bank, 2001: xi), and, *inter alia,* social capital:

> can reduce poverty through micro and macro channels by affecting the movement of information useful to the poor, and by improving growth and income redistribution at the national level (World Bank, 2001: xi).

In the UK, in government documents, social capital is seen as a mechanism for fostering community cohesion, and encouraging civic and social engagement. Redistribution of growth or income, the reduction of poverty do not figure in the use of the term in this instance. In a review of community involvement carried out for the Office of the Deputy Prime Minister, the author defines social capital as, 'attachment to place, relations with other residents' (ODPM, 2003: 6), and notes that:

> the strengthening of basic social capital underlies participation policies but is obscured by the disproportionate focus on more formal types of participation, and needs to be a policy objective in its own right (ODPM, 2003: 6).

Emphasising the participative nature of social capital, the author notes that, 'a government commitment to build social capital would not be starting from scratch. To some degree this is what community development has attempted to do for the last two generations' (ODPM, 2003: 56). Despite the difficulties of application and evaluation, it is the very breadth of the term that appeals to governments and policy makers, wishing to illustrate common problems and common solutions.

In similar fashion, some writers (Doling, 1997) have commented on a tradition of tenure ideology in Britain and the USA, where tenures are assumed to possess characteristics that affect behaviour. Home ownership is therefore propagated as offering 'ontological security' and providing a stake in a 'property-owning democracy'. In contrast, social housing is viewed as a marginalised, residual tenure, offering little opportunity for self-improvement and only leading to welfare dependency (Saunders, 1990). The problem of considering tenure in comparative terms is that it deals with the sphere of consumption and ignores production, as Ambrose (1991) has shown. In addition, tenure 'can be seen in each country to be the product of a specific interrelationship of political, economic and ideological factors' (Harloe and Maartjens, 1983: 266).

For example, home ownership in Germany is associated with settling down and family living; much housing is purpose built for the owner intending to remain in it for many years. It is therefore perceived as a minority activity pursued by just 40 per cent of households (Ball et al., 1988). In contrast, in a country like Singapore home ownership is seen as a deliberate policy to promote nation building in post-colonial state. It is viewed as a mechanism for generating high levels of household savings (Lee et al., 2003)

Dimensions of tenure take social, economic, legal, ideological and political forms. Tenure can also change over time. For example, perceptions of home ownership in Britain in the 1990s were associated with the concepts of risk and negative equity rather than security and stability. For writers such as Ball et al. (1988) the temporal and spatial specificity of tenure require a new method of conceptualising housing through the notion of 'structures of housing provision'. For others, the idea of tenure as a 'single, universal housing shorthand' cannot be considered valid and should be abandoned (Barlow and Duncan, 1988: 226). These differences illustrate that there are few fixed points for the purposes of analysis. Tenure (at the heart of most explorations of housing systems in Britain) thus differs significantly in its social construction between different countries. Many comparative studies begin with comparisons between different European countries or between Britain and Europe, but where subsidy arrangements, legal infrastructure and so on are very different the difficulty of saying anything meaningful is notable (Barlow and Duncan, 1988; 1994).

The third dimension of housing analysis; policy *transfer* relates to the practical implementation of policies within different cultural contexts. Housing policy has seen a number of initiatives which have been applied to different societies, in particular, allocation and regeneration policies.

A study of policy transfer between local regeneration partnerships in the UK noted that, 'government policy has increasingly incorporated a learning dimension into its regeneration programmes' (Wolman and Page, 2000: 2), and much of the research published by the Joseph Rowntree Foundation (JRF) and by the ODPM in Britain, relies on the notion of successful policy transfer. Indeed, the JRF launched an area regeneration programme in 1996: a research programme to assess regeneration policy and practice. The dissemination of good practice examples and the use of case studies in government reports draw heavily on the notion that 'good policy' can be transferred between organisations, often diminishing the difficulties of so doing. In fact, such transfer is highly problematic both at local and national level, and a social constructionist approach allows those problems to be more transparent.

Policy transfer in Britain has tended to be associated with American models of welfare such as 'welfare to work', marking a more punitive and highly stigmatised welfare service (Dolowitz, 2000). Such policies neglect the very different cultural settings within which US and British welfare policies can be applied. Equally one can argue that the focus on antisocial behaviour and personal responsibility in much of the UK policy is influenced by US discourses of an urban underclass. US policy is strongly influenced by a discourse of 'ghettoisation' (Burchadt et al., 2002: 2) reflecting wide disparities in income and status.

Area-based initiatives provide a specific example of where policy transfer can be argued to have been used inappropriately. In Britain, an area-based approach to poverty has characterised urban policy initiatives over a long period of time, and continues to do so. Kleinman (2000) argues that British welfare policy after 1997 was largely based on US models of the 'War on Poverty' in the early 1960s under Kennedy and Johnson. Not only are these based on assumptions that no longer hold, they also ignore more fundamental social differences.

The notion that concentrations of poverty can best be addressed by area targeting is inherent in the New Deal for Communities programme established in 1988, providing funding for 39 neighbourhoods across England with £2 billion of funding. This programme has strong similarities to the area-based approach to poverty pursued in the US, rejecting redistributionist mechanisms applied through general taxation and instead, focusing on trying to ensure that, through area-based policy interventions, 'within 10 to 20 years, no-one should be disadvantaged by where they live' (Social Exclusion Unit, 1998). The idea that targeting deprived areas can compensate for disadvantage is widely held among British politicians. Contrasting the American and European approach to residential disadvantage, Kearns notes that:

> the thesis that poor people are further disadvantaged by living in poor areas, especially to the point where their life chances and aspirations are detrimentally affected, is not as widely subscribed to in the American polity and academia (Kearns, 2002: 146).

Several studies in Britain have demonstrated that area-based intervention does not have long-term effects (Power and Tunstall, 1995; Austin Mayhead and Co., 1997; Carley and Kirk, 1998; Evans, 1998) and this may be as a result of the lack of concentration of the poor in particular neighbourhoods. The overview report of the area regeneration programme funded by the JRF concluded in 1990 that, 'while there is clear evidence of increasing concentration of deprivation, due in part to housing allocation policies and mobility of households in work, over-reliance on favoured area-based policies present challenges' (Carley et al., 2000: 3), noting that 'targeting of worst areas always misses a significant proportion of deprived households, a fact which has bedevilled urban policy for thirty years' (Carley et al., 2000: 3). In fact, in the US deprivation is more concentrated and area-based initiatives have greater relevance as the targeting of policy is more likely to be effective in tackling poverty and social deprivation.

The transfer of a policy from one country to another was notable in the adoption of 'choice-based' lettings, a policy pursued in Holland, referred to as the *Delft* model, and now introduced in Britain. The introduction of 'choice' into British housing allocations policies was signalled in the Housing Green Paper (ODPM, 2000), increasing choice through lettings policies that allow the potential tenant to choose from a number of housing options. Local authorities in Britain are now compelled to produce a plan to move towards such a scheme. However the scheme has been applied selectively and without careful attention to the context within which policy is delivered (Boelhouwer et al., 1997). In particular, the concept does not pay sufficient attention to problems of supply and demand, common to many UK social housing providers. As mentioned above, the availability of EU research grants and eagerness of policy makers to listen to solutions 'from abroad' make an increased interest in policy transfer inevitable.

Conclusion

The particularities referred to above point to the need for more awareness of the often complex social relationships and historical compromises of the past that have helped shape housing and welfare. Broad sweep generalisations need to be qualified by an understanding of the way layers of decisions and arrangements have been arrived at over time and changed as different policy ideas and constellations of power relationships have been negotiated and renegotiated. What the neglect of the key issues mentioned above implies is an inattention to the path dependencies or accumulated histories of different models of housing policies.

The value of a social constructionist approach lies in questioning the assumptions implicit in much housing discourse. This is particularly important in relation to the contemporary policy discourse of globalisation. It alerts us to other alternatives and allows a more detailed analysis of why particular problems gain currency within specific cultural settings. Whilst this chapter can only indicate some initial methodological questions, it makes the case for more detailed empirical research acknowledging some of the complexities and cultural contexts within which contemporary housing problems and solutions vie for attention.

A social constructionist approach to housing policy develops the concept of institutionally bounded limits to change. Once mechanisms of service provision have become embedded as part of a social and political structure, they are very difficult to change. The fear from politicians of electoral reaction and the power of vested organised interests is often underplayed in discussions of policy change. Thus, an emphasis on global explanations, whilst important, should not eclipse the distinctiveness of domestic policies. An over-reliance on simplifications, typologies, comparisons and classifications can obscure more obvious features of welfare systems which may be apparent to native speakers and those immersed within a cultural setting. At the same time, a sense of detachment may also be essential to gaining a sense of perspective in analysing welfare policies. Nevertheless, one of the crucial variables to consider will be the relative power of organised interest groups within the societies under study (Ellison and Pierson, 2003: 5). Gaining this sense of environmental, institutional and cultural context is crucial to a social constructionist methodology in comparative research.

The agency-structure problematic is most challenging in the study of large-scale processes of change taking place over decades, generations or even centuries. This is the *longue durée* of historical time that the theory of structuration (Giddens, 1984: 35–6) addressed, as a means of conceptualising the ways in which the innumerable everyday interactions cumulatively result in decisions that mould and remould organisational forms and that sediment institutional practices. This is the challenge facing a societal-scale application of social constructionism that Berger and Luckmann (1967) launched nearly 60 years ago. That challenge is being taken up by actor-network theory and the new institutionalism and is being applied in many different contexts. International comparative housing research can also take up that challenge and develop more context-sensitive and interaction-grounded approaches to international housing differences, while also contributing to these larger debates.

References

Abrahamsom, P. and Hansen, F. (1996), *Poverty in the European Union*, Report to the European Parliament, Roskilde, Denmark: Centre for Alternative Social Analysis, Roskilde University.

Ambrose, P. (1991), 'The Housing Provision Chain as a Comparative Analytical Framework', *Scandinavian Housing and Planning Research*, Vol. 8, No. 2: 91–104.

Austin Mayhead and Co. (1997), *Neighbourhood Renewal Assessment and Renewal Areas*, London: Department of the Environment, Transport and the Regions (DETR).

Ball, M., Harloe, M. and Martens, M. (1988), *Housing and Social Change in Europe and the USA*, London: Routledge.

Barlow, J. and Duncan, S. (1988), 'The Use and Abuse of Tenure', *Housing Studies*, Vol. 3, No. 4: 29–31.

Barlow, J. and Duncan, S. (1994), *Success and Failure in Housing Policy: European Systems Compared*, Oxford: Pergamon.

Berger, P. and Luckmann, T. (1967), *The Social Construction of Reality: a Treatise in the Sociology of Knowledge*, Garden City, New York: Anchor Books.

Blumer, H. (1969), *Symbolic Interactionism: Perspective and Method*, Englewood Cliffs, NJ: Prentice Hall.

Boelhouwer, P., van der Heijden, H. and van de Nen, B. (1997), 'Management of Social Rented Housing in Western Europe', *Housing Studies*, Vol. 12, No. 4: 509–29.

Burchardt, T., Le Grand, J. and Piachaud, D. (2002), 'Introduction', in J. Hills, J. Le Grand and D. Piauchaud (eds), *Understanding Social Exclusion*, Oxford: Oxford University Press: 1–12.

Byrne, D. (1999), *Social Exclusion*, Buckingham: Open University Press.

Carley, M. and Kirk, K. (1998), *Sustainable by 2020? A Strategic Approach to Urban Regeneration for Britain's Cities*, Bristol: Policy Press.

Carley, M., Campbell, M., Kearns, A., Wood, M. and Young, R. (2000), *Regeneration in the 21st Century: Policies into Practice – an Overview of the Joseph Rowntree Foundation Area Regeneration Programme*, Bristol: Policy Press.

Clapham, D. (1997), 'The Social Construction of Housing Management Research', *Urban Studies*, Vol. 34, Nos 5–6: 761–74.

Council of the European Union (2002), *Fight against Poverty and Social Exclusion: Common Objectives for the Second Round of National Action Plans – Endorsement*, doc. No. 14164/1/02 REV 1 SOC 508, Social Protection Committee, European Union: Brussels.

Dolowitz, D. (2000), *Policy Transfer and British Social Policy: Learning from the USA?*, Buckingham: Open University Press.

Doling, J. (1997), *Comparative Housing Policy – Government and Housing in Advanced Industrialised Countries*, London: Macmillan.

Ellison, N. and Pierson, C. (2003), 'Introduction: Developments in British Social Policy', in N. Ellison and C. Pierson (eds), *Developments in British Social Policy, 2*, London: Palgrave: 1–14.

Evans, R. (1998), *Housing Plus and Urban Regeneration: What Works, How, Why and Where?*, Liverpool: European Institute for Urban Affars, Liverpool John Moores University.

Eyraud, C. (2001), 'Social Policies in Europe and the Issues of Translation: The Social Construction of Concepts', *International Journal of Social Research Methodology*, Vol. 4, No. 4: 279–85.

Forrest, R. and Kearns, A. (2001), 'Social Cohesion, Social Capital and the Neighbourhood', *Urban Studies*, Vol. 38: 2125–43.

Franklin, B. and Clapham, D. (1997), 'The Social Construction of Housing Management', *Housing Studies*, Vol. 12, No. 1: 7–26.

Garfinkel, H. (1984 [1967]), *Studies in Ethnomethodology*, Oxford: Blackwell.

Geertz, C. (1993), *The Interpretation of Cultures*, London: Fontana.

Giddens, A. (1984), *The Constitution of Society: Outline of the Theory of Structuration*, Cambridge: Polity Press.

Granovetter, M. (1985), 'Economic Action and Social Structure: The Problem of Embeddedness', *American Journal of Sociology*, Vol. 91, No. 3: 481–510.

Granovetter, M., McGuire, P. and Schwartz, M. (1993), 'Thomas Edison and the Social Construction of the Early Electrical Industry in America', in R. Swedberg (ed.), *Explorations in Economic Sociology*, New York: Russell Sage Foundation: 213–46.

Granovetter, M. and McGuire, P. (2004), 'Shifting Boundaries and Social Construction in the Early Electricity Industry, 1878–1915', in J. Porac and M. Ventresca (eds), *Constructing Industries and Markets*, Elsevier Press.

Hallett, G. (1988), *Land and Housing Policies in Europe and the USA*, London: Routledge.

Harloe, M. (1985), *Private Rented Housing in the United States and Europe*, London: Croom Helm.

Harloe, M. (1995), *The People's Home: Social Rented Housing in Europe and America*, Oxford: Blackwell.

Harloe, M. and Maartjens, M. (1983), 'Comparative Housing Research', *Journal of Social Policy*, Vol. 13, No. 3: 255–77.

Heidenheimer, A., Heclo, H. and Adams, C. (1975), *Comparative Public Policy: The Politics of Social Choice in Europe and America*, London: Macmillan.

Hutton, W. (2002), *The World We're In*, London: Little, Brown.

Jacobs, K. and Manzi, T. (2000), 'Evaluating the Social Constructionist Paradigm in Housing Research', *Housing, Theory and Society*, Vol. 17, No. 1: 35–42.

Jacobs, K., Kemeny, J. and Manzi, T. (2003), 'Power, Discursive Space and Institutional Practices in the Construction of Housing Problems', *Housing Studies*, Vol. 18, No. 4: 429–46.

Johnston, G. and Percy-Smith, J. (2003), 'In Search of Social Capital', *Policy and Politics*, Vol. 31, No. 3: 321–34.

Jorgensen, D.L. (1989), *Participant Observation: A Methodology for Human Studies*, London: Sage.

Karn, V. and Wolman, H. (1992), *Comparing Housing Systems: Housing Performance in the United States and Britain*, Oxford: Clarendon Press.

Kearns, A. (2002), 'Response: From Residential Disadvantage to Opportunity? Reflections on British and European Policy and Research', in *Housing Studies*, Vol. 17, No. 1: 145–50.

Kemeny, J. and Lowe, S. (1998), 'Schools of Comparative Housing Research: From Convergence to Divergence', *Housing Studies*, Vol. 13, No. 2: 161–96.

Kleinman, M. (2000), 'Include Me Out? The New Politics of Place and Poverty', *Policy Studies*, Vol. 21, No. 1: 49–61.

Latour, B. (1986), *Laboratory Life: The Construction of Scientific Facts*, New Jersey: Princeton University Press.

Lee, J., Forrest, R. and Tam, W. (2003), 'Home Ownership in East and South East Asia: Market State and Institutions', in R. Forrest and J. Lee (eds), *Housing and Social Change: East-West Perspectives*, London: Routledge.

Levitas, R. (1996), 'The Concept of Social Exclusion and the New Durkheimian Hegemony', *Critical Social Policy*, Vol. 46: 5–20.

Office of the Deputy Prime Minister (ODPM) (2000), *Quality and Choice in Housing: A Decent Home for All: The Housing Green Paper*, Norwich: The Stationery Office.

Office of the Deputy Prime Minister (ODPM) (2003), *Searching for Solid Foundations: Community Involvement and Urban Policy*, Chanan, G. [of] Community Development Foundation for ODPM, London: ODPM.

Pierson, P. (1994), *Dismantling the Welfare State? Reagan, Thatcher and the Politics of Retrenchment*, Cambridge: Cambridge University Press.

Power, A. and Tunstall, R. (1995), *Swimming Against the Tide: Polarisation or Progress on 20 Unpopular Council Estates, 1980–1995*, York: Joseph Rowntree Foundation.

Putnam, R. (2001), *Bowling Alone: the Collapse and Revival of American Community*, London: Simon and Schuster.

Russell Bernard, H. (1998), *Research Methods in Anthropology*, London: Sage.

Saunders, P. (1990), *A Nation of Home Owners*, London: Unwin Hyman.

Schutz, A. (1962), *Collected Papers 1: The Problem of Social Reality*, The Hague: Martinus Nijhoff.

Silver, H. (1995), 'Reconceptualising Social Disadvantage: Three Paradigms of Social Exclusion', in G. Rodgers, C. Gore and J. Figeuiredo (eds), *Social Exclusion: Rhetoric, Reality, Responses*, Geneva: ILO.

Social Exclusion Unit (SEU) (1997), *Bringing Britain Together: A Strategy for National Renewal*, HMSO, CM 4045.

Somerville, P. and Bengtsson, B. (2002), 'Constructionism, Realism and Housing Theory', *Housing, Theory and Society*, Vol. 19, Nos 3–4: 121–36.

Stébé, J.-M. (1999), *La Crise des Banlieus, Que Sais-Je?*, No. 3507, Paris: Presses, Universitaires de France.

Strauss, A. (1978), *Negotiations: Varieties, Contexts, Processes and Social Order*, San Francisco: Jossey-Bass.

Tizot, J. (2001), 'The Issues of Translation, Transferability and Transfer of social Policies: French and British "Urban Social Policy": Finding Common Ground for Comparison?', *International Journal of Social Research Methodology*, Vol. 4, No. 4: 301–17.

Whyte, W. (1993 [1955]), *Street Corner Society, the Social Structure of an Italian Slum*, Chicago: Chicago University Press.

Wolman, H. and Page, E.C. (2000), *Learning from the Experience of Others: Policy Transfer among Local Regeneration Partnerships*, York: Joseph Rowntree Foundation.

World Bank (2001), 'Understanding and Measuring Social Capital: A Synthesis of Findings and Recommendations from the Social Capital Initiative', Social Capital Initiative weekly paper No. 24, Social Development Family, Environmentally and Social Sustainable Development Network, <http://www.worldbank.org/social development>.

Zukin, S. and Di Maggio, P. (1990), *Structures of Capital: The Social Organisation of the Economy*, Cambridge: Cambridge University Press.

Index